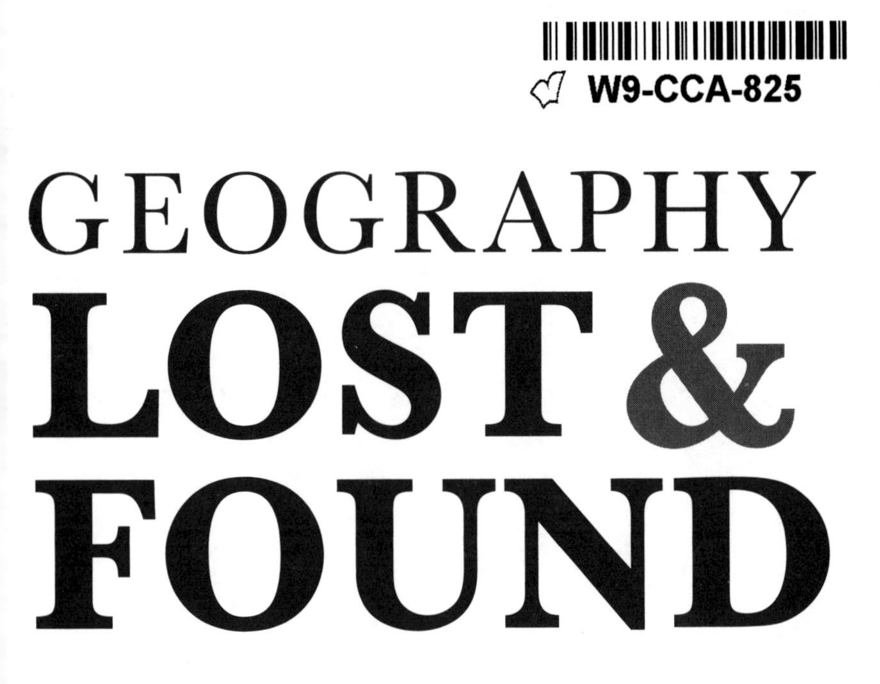

Publications International, Ltd.

Cover art: Art Explosion, Getty, Shutterstock.com

Interior art: Art Explosion, Getty, Shutterstock.com

Louis Weber, CEO
Publications International, Ltd.
7373 North Cicero Avenue
Lincolnwood, Illinois 60712

ISBN: 978-1-68022-127-5

Manufactured in Canada.

8 7 6 5 4 3 2 1

TABLE OF CONTENTS

INTRODUCTION

Welcome to *Geography: Lost and Found*. In this book, you'll find more than 150 funny, informative articles that will take you on a trip through time and around the world. Read about ancient cities, the origins of cultural traditions, and weird quirks of our world today. Discover fascinating facets of our natural world as you read about everything from flesh-eating plants to poisonous animals. Travel from the North Pole to the depths of the sea. Along the way, you'll get answers to questions that will intrigue you, pique your interest, and tickle your funny bone. These are just a few of the questions you'll find answered:

How long a bird can fly without stopping?

Did the French really invent French toast?

Are good manners the same throughout the world?

Take a few hours to lose yourself in this book. You'll love what you find!

HERE, THERE, AND EVERYWHERE

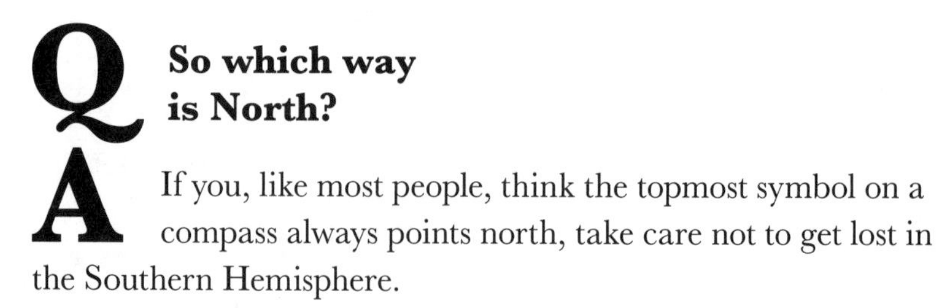

Q **So which way is North?**

A If you, like most people, think the topmost symbol on a compass always points north, take care not to get lost in the Southern Hemisphere.

North, south, east, and west: These directions are meant to be set in stone, the unchangeable points of reference that lead sailors through treacherous seas and intrepid adventurers through dark and unknown lands. Yet even these directions, such stalwarts of clarity and precision, come with a medley of misconceptions.

One misconception is that a compass points to the North Pole. First, we need to clarify what we mean when we talk about the North Pole. We could be talking about four different locales: geographic, magnetic, geomagnetic, and the pole of inaccessibility.

The geographic North Pole, known as true north or ninety degrees north, is where all longitudinal lines converge. It sits roughly four hundred and fifty miles north of Greenland, in the center of the Arctic Ocean.

The magnetic pole—the point marker for compasses—is located about one hundred miles south of the geographic pole, northwest of the Queen Elizabeth Islands, which are part of northern Canada. Earth has a magnetic field, which is created by the swirling motion of molten lava that resides in its core. This magnetic field makes an angle with Earth's spin axis. The geographic poles, in contrast, are the places that Earth's imaginary spin axis pass straight through. So while the geographic and magnetic poles are close to each other, they are never in the exact same place. If you're heading "due north" as the compass reads, you're heading to the North Magnetic Pole, not the North Pole. But compasses don't work close to a magnetic pole, so if you're going to the North Pole, a compass will take you only so far.

To complicate matters further, the North Magnetic Pole is always moving, because the motion of the swirling lava changes. In 2005, the North Magnetic Pole was 503 miles from the geographic North Pole, placing it firmly in the Arctic Ocean, north of Canada. Meanwhile, the South Magnetic Pole was 1,756 miles from the geographic South Pole, in Antarctica just south of Australia. The rates of change of the magnetic poles vary, but lately they've been moving at approximately 25 miles per year. Scientists project that in 50 years, the North Magnetic Pole will be in Siberia.

Then there's the North Geomagnetic Pole, the northern end of the axis of the magnetosphere, the geomagnetic field that surrounds the earth and extends into space. Last is the Northern Pole of Inaccessibility, the point in the Arctic Ocean that is most distant from any landmass.

That moves us right along to misconception number two, which is that the "N" on a compass always points north. Assuming that the designation "north" always coincides with the notion of "up," the runaway North Magnetic Pole reminds us that "up" is relative. Earth is a sphere, so logically any single point could be designated as the top, making whatever lies opposite this arbitrary top the arbitrary bottom. It made sense for early mapmakers to draw the North Pole at the top of a map, because this was their approximation of where that handy compass pointed. As the North Magnetic Pole continues to wander from the point cartographers deemed the "North Pole," the designation of this geographic location as due north may eventually become obsolete.

Meanwhile, in the Southern Hemisphere, a compass points toward the South Magnetic Pole and more or less toward the corresponding yet inevitably inaccurate location that is deemed the geographic South Pole. Early mapmakers hailed from the Northern Hemisphere, so the North Pole is logically represented as being at the top of the map and the top of the world. Yet in the Southern Hemisphere, it would be equally logical to place the South Pole at the top of the map.

Q How many people live at the North Pole besides Santa and his elves?

A The population at the North Pole is as transient as the terrain itself, which is in a constant state of flux due to shifting and melting ice. Human life in this frigid region consists of researchers floating on makeshift stations and tourists who aren't the sit-on-a-beach-in-the-Bahamas type. There are no permanent residents at the North Pole—save, of course, for Santa Claus and his posse.

If the North Pole were more like its counterpart on the other end of the earth, the South Pole, it would be a lot more accessible. Since the South Pole is located on a continent, Antarctica, permanent settlements can be established. In fact, research stations at the South Pole have been in place since 1956. These bases range in population size, but most average fifteen personnel in winter (April to November) and one hundred and fifty in summer (December to March). Combined, the stations house a few thousand people in the summer. The U.S. McMurdo Station alone might exceed a thousand individuals at the peak time of year.

All of this helps explain why Santa chose to live at the North Pole rather than the South Pole. If you're S. Claus and you don't want to be found, there isn't a better place than the North Pole to set up shop, even if that shop is always in danger of floating away.

Q Does anyone send a telegram anymore?

A On May 24, 1844, Samuel Morse, inventor of the telegraph and of Morse code, sent this message over an experimental line from Washington, D.C., to Baltimore: "What hath God wrought?" That was the first telegram, and more than 160 years later, young fingers echo the sentiment with every "OMG!" that's punched into a smartphone keyboard. It's an appropriate, albeit unintended, homage—for what was a telegram if not Victorian texting?

Decades before any of today's forms of instantaneous communication existed, the telegram was the quickest way to bother somebody who wasn't within earshot. You'd visit a telegraph office and dictate a message to a telegraph operator. The operator would tap it out on a "key" that transmitted the message over the telegraph wire in Morse code, a series of long and short electric pulses that represent letters and numbers. An operator on the receiving end would then transcribe the code into a printed message.

And like magic, your thoughts would reach your intended target within a matter of hours, delivered in person by a uniformed messenger.

This was revolutionary technology at a time when long-distance communication relied on the mail. And it remained popular even after the advent of the telephone. Western Union, the company most identified with the telegram, sent more than two hundred million of them in 1929. People used them to convey news of births, deaths, and marriages; some pleaded for money, others offered congratulations. The printed messages often were saved for years as markers of life's big events.

But the world has changed considerably since then. In this day of email, smartphones, video conferencing, and other technologies that keep us wired in, does anybody still send a telegram?

Bowing to dwindling demand, Western Union stopped telegram service in January 2006. Western Union delivered only about twenty thousand telegrams in its final year.

But the telegram isn't dead. There are still businesses that will deliver one for you, and there are a few remote places in the world where you can stumble across a telegraph office, such as small Balkan villages in Eastern Europe, where electricity and phones are relatively scarce. And you can be sure that people are still asking for money.

Do countries get paid for delivering mail from other countries?

Yes, but fortunately, countries work out the tab amongst themselves. Imagine having to buy stamps from every country that handled your letter as it went around the world. That was how people did it in the early days of mail, and it was a real pain.

In 1874, a bunch of nations got together and formed an international organization to sort it all out. The goals were to eliminate the need for countries to establish individual postal treaties with one another and to allow people to buy stamps from whatever country they were in. The organization is now a United Nations agency called the Universal Postal Union (UPO).

Over the years, the UPO has adjusted its formula to make certain that every country receives a fair share of the dough. Initially, the UPO assumed that almost every letter would get a reply, meaning that any two countries would spend about the same amount of time and money delivering mail from the other. As a result, participating countries kept all the money for mail leaving their shores. But with the rise of magazine delivery, mail-order business, and the like, some countries (including the United States) ended up getting the short end of the stick, receiving more mail than they sent. So the UPO instituted terminal dues: payments from the country of origin to the destination country to cover the costs associated with foreign mail.

Today, terminal dues are based on a complex formula that factors in the total weight of the mail and the total number of pieces going from one nation to another, as well as the quality of service in the destination country. For industrialized countries, the formula takes into account the cost of delivering mail in the destination country. For developing countries, the formula uses an average world rate instead of an individual rate.

The math is complicated, and the UPO seems to be forever tweaking its formula. Just be glad the United Nations figures it all out so that you don't have to.

Q Why do crows always take the most direct route?

A Those large, perching black birds are pretty darn smart, maybe even smarter than your toddler. Can your kid count aloud up to seven, say more than one hundred words, and speak in complete sentences? Well, some highly intelligent crows can—they mimic the human voice, much like a parrot does.

Now, as for being smart enough to know the most direct route, perhaps you're thinking of the expression "as the crow flies." That means traveling in a straight line, or taking the fastest or shortest route from point A to point B. Some think this idiom is based on the notion that crows, being very shrewd, always fly straight to the nearest food supply.

Ornithologists will tell you that some crows—such as the American Crow—do indeed fly with very deliberate, flapping wing beats. However, if you take a look out the window, you might notice your neighborhood crows aren't necessarily flying on the straight and narrow. In fact, in their forage for food, they're probably swirling around in patterns similar to large, wheeling arcs.

So what gives? Well, "as the crow flies" is actually a very old expression. The earliest known printed use of the phrase came in 1767, when William Kenrick wrote this in *The London Review of English and Foreign Literature:* "The Spaniaad [sic], if on foot, always travels as the crow flies, which the openness and dryness of the country permits; neither rivers nor the steepest mountains stop his course, he swims over the one and scales the other."

Need a translation? Kenrick alludes to the fact that crows (and apparently Spaniards) are not encumbered by obstacles of land, road, or sea. For humans, though, rivers, mountains, and construction detours on U.S. Route 20 often force us to take deviating and more time-consuming paths.

Case in point: The distance between Key West, Florida, and Pensacola, Florida, is 524 miles directly across the Gulf of Mexico, as the crow flies. But if you follow Google Maps and take that trip by car, you'll be riding all the way up the Sunshine State for about fourteen hours, or a distance of about 800 miles.

So it seems the idea that crows always take the most direct route has more to do with colloquial phraseology than literal fact. Truth is, any flying bird is able to get between two points without being hindered by roadblocks. But that doesn't mean you should sell those crafty crows short.

It turns out these bright birdies were quite the feathered friends to early sailors. According the U.S. Navy's *Origin of Navy Terminology,* crows were once carried onto ships as a sort of onboard GPS system. In cases of poor visibility, a crow was released into the air.

Why? Crows are landlubbers, so they inevitably headed straight toward dry land. This allowed the navigator of the ship to plot a course to shore, even in foggy weather. Now just where were these fowl crewmembers kept? In a cage, high up on the ship's main mast: the "crow's nest," of course!

Q How do carrier pigeons know where to go?

A No family vacation would be complete without at least one episode of Dad grimly staring straight ahead, gripping the steering wheel and declaring that he is not lost as Mom insists on stopping for directions. Meanwhile, the kids are tired, night is falling, and nobody's eaten anything except a handful of Cheetos for the past six hours. But Dad is not lost. He will not stop.

It's well known that men believe they have some sort of innate directional ability—and why not? If a creature as dull and dimwitted as a carrier pigeon can find its way home without any maps or directions from gas-station attendants, a healthy human male should certainly be able to do the same.

Little does Dad know that the carrier pigeon has a secret weapon. It's called magnetite, and its relatively recent discovery in the beaks of carrier pigeons may help solve the centuries-old mystery of just how carrier pigeons know their way home.

Since the fifth century BC, when carrier pigeons were used for communication between Syria and Persia, they have been prized for their ability to find their way home, sometimes over distances of more than five hundred miles. In World War I and World War II, Allied forces made heavy use of carrier pigeons, sending messages with them from base to base to avoid having radio signals intercepted or if the terrain prevented a clear signal. In fact, several carrier pigeons were honored with war medals.

For a long time, there was no solid evidence to explain how these birds were able to find their way anywhere, despite theories that ranged from an uncanny astronomical sense to a heightened olfactory ability to an exceptional sense of hearing. Then scientists made an important discovery: bits of magnetic crystal, called magnetite, embedded in the beaks of carrier pigeons. This has led some researchers to believe that carrier pigeons have magneto reception—the ability to detect changes in the earth's magnetic fields—which is a sort of built-in compass that guides these birds to their destinations.

Scientists verified the important role of magnetite through a study that examined the effects of magnetic fields on the birds' homing ability. When the scientists blocked the birds' magnetic ability by attaching small magnets to their beaks, the pigeons' ability to orient themselves plummeted by almost 50 percent. There was no report, however, on whether this handicap stopped male pigeons from plunging blindly forward. We'd guess not.

Q Why is Australia considered a continent instead of an island?

A In grammar school, some of us were far more interested in the "social" aspect of social studies than the "studies" part. Nevertheless, everyone can recite the continents: Africa, Asia, Europe, South America, North America, Australia, and...some other one.

What gives with Australia? Why is it a continent? Shouldn't it be an island?

It most certainly is an island (the world's largest) and so much more. Australia is the only land mass on Earth to be considered an island, a country, and a continent.

Australia is by far the smallest continent, leading one to wonder why it is labeled a continent at all when other large islands, such as Greenland, are not. The answer lies in plate tectonics, the geologic theory explaining how Earth's land masses got to where they are today. According to plate tectonic theory, all of Earth's continents once formed a giant land mass known as Pangaea. Though Pangaea was one mass, it actually comprised several distinct pieces of land known as plates.

Over millions of years, at roughly the speed of your hair growth, these plates shifted, drifting apart from one another until they reached their current positions. Some plates stayed connected, such as South America and North America, while others moved off into a remote corner like a punished child, such as Australia. (It's no wonder Australia was first used by the British as a prison colony.) Because Australia is one of these plates—while Greenland is part of the North American plate—it gets the honor of being called a continent.

All of this debate might ultimately seem rather silly. Some geologists maintain that in 250 million years, the continents will move back into one large mass called Pangaea Ultima. Australia will merge with Southeast Asia—and social studies tests will get a whole lot easier.

Q Why does it take longer to fly west than east?

A If you've ever flown a long distance, you might have noticed that it takes more time to fly from east to west than west to east. What's the deal with that? Since there's no real traffic in the sky (apart from the occasional flock of birds), the delay seems inexplicable. It should take the same amount of time, regardless of direction, right?

But the bitter truth is, the air up there isn't quite as wide open as it seems. At the high altitude that is required for commercial flights, there's a powerful and persistent horizontal wind known as the jet stream. Because of the differences in temperature and pressure between the equator and the earth's polar regions, the jet stream flows from west to east in the Northern Hemisphere.

This jet stream is a lot like a current in a river: If you're moving with the current, you'll go faster; conversely, working against the current slows you down and makes you work harder to get where you're going.

The airlines, of course, know all about the wily ways of the jet stream. They take advantage of it, purposely flying within it on eastbound flights to allow planes to reach their destinations sooner and with less wear and tear. (Some airlines even offer cheaper fares on the eastbound leg of a journey.) But on a westbound flight, a pilot must fly against the jet stream, which obviously means that it takes more time.

A westerly cross-country flight lasts about a half-hour longer than its easterly counterpart. Though it's invisible to the naked eye, the jet stream is like a giant traffic jam in the sky.

Q Do birds get tired in flight?

A Flight—especially migration—can be an exhausting experience for any bird. Reducing the amount of energy that is spent in the air is the primary purpose of a bird's body structure and flight patterns. Even so, migrating thousands of miles twice a year takes its toll on a bird's body, causing some to lose up to 25 percent of their body weight. How do they keep on truckin'?

Large birds cut energy costs by soaring on thermal air currents that serve to both propel them and keep them aloft, which minimizes the number of times the beasts have to flap their wings. The concept is similar to a moving walkway at an airport: The movement of the current aids birds in making a long voyage faster while expending less energy.

Smaller birds lack the wingspans to take advantage of these currents, but there are other ways for them to avoid fatigue. The thrush, for instance, has thin, pointed wings that are designed to take it great distances while cutting down on the energy expended by flapping. Such small birds also have light, hollow bone structures that keep their body weights relatively low.

If you've ever seen a gaggle of migrating geese, you likely noticed the distinctive V-formation that they take in flight. They do this to save energy. The foremost goose takes the brunt of the wind resistance, while the geese behind it in the lines travel in the comparatively calm air of the leader's wake. Over the course of a migration, these birds rotate in and out of the leader position, thereby dispersing the stress and exhaustion.

While large birds routinely migrate across oceans, smaller birds tend to keep their flight paths over land—they avoid large bodies of water, mountain ranges, and deserts. This enables them to make the occasional pit stop.

Perhaps the most amazing avian adaptation is the ability to take short in-flight naps. A bird accomplishes this by means of unilateral eye closure, which allows it to rest half of its brain while the other half remains conscious. In 2006, a study of Swainson's Thrush—a species native to Canada and some parts of the United States—showed that the birds took hundreds of in-flight naps a day. Each snooze lasted no more than a few seconds, but in total, they provided the necessary rest.

How sweet is that? After all, who among us wouldn't want to take naps at work while still appearing productive?

Q How far can a bird fly without stopping?

A A female bar-tailed godwit named E7 set the record in 2007. This topflight bird, which was fitted with a satellite transmitter, left the shores of Alaska and proceeded to fly 7,145 miles across the Pacific Ocean to New Zealand (at an average speed of about 35 miles per hour).

Unlike other long-distance flyers—such as the albatross or the Arctic tern, which stop on occasion to feed or rest on the ocean's surface—E7 plowed ahead without the benefit of food, water, or rest and completed the journey in a little more than eight days.

This achievement, which was measured by United States Geological Survey scientists, ranks as the longest interrupted bird migration on record. And to think—this Homeric traveler weighs less than a pound.

Traversing long distances without stopping requires the right type of aircraft, and birds in general have ideal equipment. Their anatomies are engineered for flight efficiency, from their feather patterns and wing designs to their hollow skeletons and finely tuned digestive systems, which can shut down to conserve energy. Even their air intake allows oxygen to quickly flow through in one direction.

Birds also have a navigation device of sorts: pressure-sensitive ears that predict altitude and weather fronts. Additionally, their ability to glide, soar, hover, and—in cases of migration—use air stream currents that flow off of other birds within their V-formations enables them to stay aloft without actively using their wings.

No flying feat demonstrates the wonders of these aerodynamic features better than little E7's journey. Her record-setting trek adds new meaning to the term "intelligent design."

Q Are there places where it's too cold to start a fire?

A With the right materials and tools, you can start a fire anytime and anywhere. Fire is simply one result of a chemical reaction between an oxidizer (typically oxygen in the atmosphere) and some sort of fuel (for example, wood or gasoline).

To trigger this reaction, you need to excite the chemicals in the fuel to the point that they will break free and combine with oxygen in the air to form new chemical compounds. In other words, you need to heat the fuel to its ignition point. Once you get the reaction going, the atoms involved will emit a lot of heat, to the point that they glow, producing flames. If the flames from the reaction are hot enough, they will heat more fuel to its ignition point and the fire will spread. The fire will keep itself going until it runs out of fuel.

Cold air does make this process more difficult. Heat energy in the fuel dissipates in the surrounding air, which cools the fuel. The colder the air temperature, the more quickly the fuel will cool. As a result, it takes more energy to heat the fuel to its ignition point in the cold. The two opposing processes go head-to-head: Air is working to cool the fuel while the ignition source is working to heat it up. Whichever process acts more quickly will prevail.

Different types of fuel produce different degrees of heat, which determine their ability to outpace the cooling effect of cold air. For example, propane gas burns much hotter than an equal amount of wood. Heat also increases depending on how much fuel is burning—a burning log cabin is a lot hotter than a single burning log.

Say you are in Antarctica, and it is –129 degrees Fahrenheit (the lowest natural temperature ever recorded on the planet)—you would not be able to get a campfire going by rubbing two sticks together. The cold air would cool the sticks faster than you could heat them to the ignition temperature of wood (four hundred to five hundred degrees Fahrenheit). However, even at that low temperature, you could heat a tank of propane with an electric burner and get a fire going with a spark. You could then use that flame to start a massive wood fire—the collective heat of all the burning wood would outpace the cooling effect of the cold air.

As long as you take along some oxygen, you can even get a fire going in space, where there's no atmosphere to warm things up. Spacecraft use rockets that combine fuel and a compound that contains oxygen to trigger a combustion reaction; the fuel burns, releasing hot gases that push the rocket forward. So with the right equipment, you could theoretically get a campfire going on the moon. You might even be able to roast some marshmallows.

Why do toilets flush backwards in Australia?

What do you mean by backwards? Toilets in the United States, we are grateful to note, flush in a general downward direction. And given the relative popularity of Australia as a travel destination, we can only infer that the toilets there do likewise. Because if they flushed in an upward direction, nobody would want to (ahem) "go" there.

If, on the other hand, you're referring to the direction in which the water in a flushing toilet swirls, the answer gets a little more complicated—though not by much.

Our fascination with Australian plumbing is related to our awareness of something called the Coriolis force. We heard about it during our high school physical science classes, but most of us paid only enough attention to remember the term. We're not so adept at recalling the details of what it is or what it does.

The Coriolis force is named for the man who developed the concept, French scientist Gaspard-Gustave de Coriolis. He did a bunch of math that provided an answer for why weather systems spin in different directions depending on whether they're north or south of the equator. Let's eschew the elaborate equations and boil it down to this: As the earth spins on its axis, large air masses are pushed along in the direction of the rotation, and the air nearest the equator moves the fastest. That, in turn, affects the spin of cyclones. So in the Northern Hemisphere, cyclones spin counterclockwise; in the Southern Hemisphere, they spin clockwise. Spin a globe (a real globe, not something Google has created) and it's pretty easy to imagine how that motion affects the atmosphere.

Some of our less-studious peers, who are armed with dangerously little knowledge of the Coriolis force, have surmised that the rotation of the earth must also have an effect on water as it swirls down a drain. But as you've probably guessed by now, they're wrong. The Coriolis effect is only demonstrable on a huge scale over a long period of time. Depending on what you've been eating lately, you may occasionally think that your flushes happen on a huge scale over a long period of time, but they're not large or long enough to qualify for Coriolis status.

The truth is that regardless of whether you're in Australia, Austria, or Austin, Texas, the direction that your toilet flushes is dictated by the design of your toilet: Some send their water in a clockwise direction, and others send it counterclockwise. And if you're lucky (and regular), it should all travel in the most important direction of all: down.

Q Why are you colder on a mountain, even though you're closer to the sun?

A This question assumes you might think that the only factor influencing temperature is proximity to that fireball in the sky. But as you doubtless know already, it's much more complicated than proximity. It has to be, or else the lowest nighttime temperatures on Mercury, which is two-and-a-half times closer to the sun than Earth is, wouldn't be –297 degrees Fahrenheit. And the moon, which is sometimes 240,000 miles closer to the sun than Earth is, wouldn't get as cold as –280 degrees Fahrenheit.

The reason it's colder on a mountaintop is because at that altitude, the atmosphere is different. Specifically, the air pressure is lower. The pressure at the top of Mount Everest, which is five-and-a-half miles above sea level, is less than a third of what it is at sea level.

During July—the warmest month on Everest—the average temperature hovers around –2 degrees Fahrenheit. It doesn't get above freezing up there, ever.

In simplest terms, when air is put under more pressure, it gets warmer. When the pressure lessens, it gets colder. That's why a bicycle pump warms up when you pump up a tire—in addition to the friction that's caused by the piston inside, the pump creates air pressure. It's also why an aerosol can gets downright cold if you spray it too long—air pressure escapes from the can.

When you think about all this, you realize how many scientific factors combine to make Earth hospitable for life. The moon has no atmosphere whatsoever; Mercury has a minute amount of it. The precise combination of gases that make up our atmosphere accounts for the air we breathe, the way the sun warms it, the color of the sky, and myriad other factors that explain life as we know it.

In fact, the cold temperatures at the top of a mountain are pretty minor when it comes to the miraculous—but scientifically logical—realities that are related to atmosphere.

Q Why can Sherpas exist at higher altitudes than anyone else?

A In the early twentieth century, when Westerners first began to dream of reaching the summit of Mount Everest, one lesson quickly became clear: It was fruitless to try to make the climb without a Sherpa.

For the average person, the atmosphere way up there—29,035 feet at the mountain's peak—is not all that different from being in outer space. The thin air and extreme cold make it a deadly environment. Hypoxia, also known as mountain sickness, sets in quickly, resulting in hallucinations and impaired judgment. (There's an argument to be made that the mere decision to attempt such a climb might be ample evidence that one's judgment is already impaired.)

But Western mountaineers on those early expeditions noticed that their Sherpa guides often seemed impervious to the dangers of high altitude. They maintained their strength and breathed the thin air with ease. The cold didn't seem to bother them, either. All in all, they were downright cheerful, even under the most dreadful conditions.

So Sherpas became indispensable climbing companions. Sir Edmund Hillary, a native of New Zealand, wasn't alone when he became the first person ever to reach the summit of Everest in 1953—Tenzing Norgay, his Sherpa guide, was right behind him, "taking photographs and eating mint cake," as Norgay later described it.

In 1963, after the first American expedition reached the summit, three climbers were unable to complete the descent of the mountain because of frostbite, and they had to be rescued by Sherpas. Teams of four Sherpas carried each man for two days. The climbers later reported that by the end of the first day, their Sherpa rescuers not only were unaffected, but they had even become competitive, racing each other back to base camp.

What makes Sherpas so special? The Sherpa people are a small ethnic group concentrated in the Himalayan regions of Nepal, India, and Tibet. About ten thousand of them live in the Khumbu Valley of Mount Everest at elevations of about ten to twelve thousand feet. The reasons for their resistance to the dangerous effects of high altitude remain a mystery. Some researchers believe that living in high-altitude villages for hundreds of years has created an inherent genetic predisposition in Sherpas that allows them to cope with the rarefied air.

Whatever the reason, the message is obvious: Always take a Sherpa with you on the way up Everest, and be sure to pack enough mint cake for everybody.

Q What would you encounter if you tried to dig a hole to China?

A Hopefully you would encounter a chiropractor, because severe back pain is about all that your journey would yield. It's obviously impossible to dig a hole to China; but for the sake of argument, we'll entertain this little gem.

Before starting, let's establish that the starting point for our hole is in the United States, where this expression appears to have originated. Nineteenth-century writer/philosopher Henry David Thoreau told the story of a crazy acquaintance who attempted to dig his way to China, and the idea apparently stuck in the American popular mind.

We also need to clear up a common misconception. On a flat map, China appears to be exactly opposite the United States. However, about five hundred years or so ago, humanity established that the earth is round, so we should know not to trust the flat representation. If you attempted to dig a hole straight down from the United States, your journey—about eight thousand miles in all—would actually end somewhere in the Indian Ocean. Therefore, our hole will run diagonally; this will have the added benefit of sparing us from having to dig through some of the really nasty parts of the earth's interior.

Anyway, let's dig. The hole starts with the crust, the outer layer of the planet that we see every day. The earth's crust is anywhere from about three to twenty-five miles thick, depending on where you are. By the time we jackhammer through this layer, the temperature will be about sixteen hundred degrees Fahrenheit—hot enough to fry us in an instant. But we digress.

The second layer of the earth is the mantle. The rock here is believed to be slightly softer than that of the crust because of the intense heat and pressure. The temperature at the mantle can exceed four thousand degrees Fahrenheit, but who's counting?

Since our hole is diagonal, we'll probably miss the earth's core. At most, we'll only have to contend with the core's outermost layer. And it's a good thing, too: Whereas the outer core is thought to be liquid, the inner core, which is about four thousand miles from the earth's surface, is believed to be made of iron and nickel, and is extremely difficult to pierce, particularly with a shovel. But either way, it would be hotter than hot; scientists think the outer core and inner core are seven thousand and nine thousand degrees Fahrenheit, respectively.

And you thought hot wings night at your local bar took a toll on your body! No, unless fire and brimstone are your thing, the only journey you'll want to take through the center of the earth is a hypothetical one.

Q Who were Rand and McNally?

A It may be hard to believe that Rand McNally, a company known for its highway maps, actually got its start before the automobile. This publishing giant dates its inception to 1856—a time when the burgeoning rail industry was hailed as the future of the United States. Bostonian William Rand knew opportunity when he saw it; he packed his bags and headed to Chicago, an area destined to become a railroad hub due to its prime location on the shores of Lake Michigan. Rand set up a small printing shop and got to work producing rail tickets.

Two years later, Rand hired Irish immigrant Andrew McNally, an experienced printer. By 1868, the men had become partners, establishing Rand McNally & Company and assuming control over the *Chicago Tribune*'s printing shop. Business increased as the pair began printing railway timetables along with the tickets.

In 1871, disaster struck Chicago as the Great Fire swept the downtown area. With flames racing toward the plant, Rand and McNally had the foresight to bury two printing presses in the sand along Lake Michigan. A mere three days later, the men had the presses up and running in a newly rented shop. This little nugget of company lore remains a source of pride more than a century later.

By 1872, Rand McNally's future success was on the map—literally. Using an innovative new printing technology that relied on wax engraving, Rand McNally & Company included their very first map in a railroad guide. The wax engraving process allowed the company to mass-produce maps in a cost-effective fashion. By 1880, Rand McNally employed 250 workers and had become the largest producer of maps in the United States. Rand McNally's reputation as a premier mapmaker allowed the company to branch out into the educational publishing market with globes, atlases, and yearly geography textbooks that included pictures, diagrams, and, of course, maps.

Rand retired at the turn of the century, and McNally died of pneumonia in 1904. McNally's descendants continued to run the firm successfully. Not surprisingly, Henry Ford's mass-produced automobiles proved to be a boon to the mapmaking business. In 1907, Rand McNally began producing its popular *Photo Auto Guides,* which added arrow overlays to maps to help automobile owners find their way around—perhaps marking the last time in history that American men would actually seek out directions when driving. Renowned aviator Charles Lindbergh boosted sales further when he revealed that he used Rand McNally railroad maps to navigate.

In the 1940s, the company published the work of Thor Heyerdahl, a maverick anthropologist and geographer who believed that the Polynesian people descended from ancient South Americans who had sailed to the distant South Pacific islands. The scientific community largely rejected Heyerdahl's theories, but when Rand McNally published *Kon-Tiki: Across the Pacific by Raft,* the book sold more than a million copies in its first few years. It is still in print today.

Over the years, maps and atlases have remained at the center of the company's business, but the firm has kept pace with the times by offering trip-planning services online. Although the McNally family sold their interest in the company in the 1990s, the firm is still based in the Chicago area and remains one of the world's most respected cartography firms.

FYI, the Rand McNally *Road Atlas* we all know and love (and have jammed under the front seat of our cars) got its start in 1960. Did someone say road trip?

Q How was the ancient city of Angkor lost and found again?

A There's some question about whether the ancient city of Angkor in Cambodia was really lost—after all, for hundreds of years, rumors of the lost city of Angkor spread among Cambodian peasants. However, on a stifling day in 1860, French botanist and explorer Henri Mahout and his porters discovered that the ancient city was more than mere legend. He gazed down weed-ridden avenues at massive towers and stone temples wreathed with carvings of gods, kings, and battles. The ruins before him were none other than the temples of Angkor Wat.

Although often credited with the discovery of Angkor Wat, Mahout was not the first Westerner to encounter the site. He did, however, bring the "lost" city to the attention of the European public when his travel journals were published in 1868. He wrote: "One of these temples—a rival to that of Solomon, and erected by some ancient Michelangelo—might take an honorable place beside our most beautiful buildings."

Mahout's descriptions of this "new," massive, unexplored Hindu temple sent a jolt of lightning through Western academic circles.

Explorers from western Europe combed the jungles of northern Cambodia in an attempt to explain the meaning and origin of the mysterious lost shrine.

Scholars first theorized that Angkor Wat and other ancient temples in present-day Cambodia were about 2,000 years old. However, as they began to decipher the Sanskrit inscriptions, they found that the temples had been erected during the 9th through 12th centuries. While Europe languished in the Dark Ages, the Khmer Empire of Indochina was reaching its zenith.

The earliest records of the Khmer people date back to the middle of the 6th century. They migrated from southern China and Tibet and settled in what is now Cambodia. The early Khmer retained many Indian influences from the West—they were Hindus, and their architecture evolved from Indian methods of building.

In the early 9th century, King Jayavarman II laid claim to an independent kingdom called Kambuja. He established his capital in the Angkor area some 190 miles north of the modern Cambodian capital of Phnom Penh. Jayavarman II also introduced the cult of devaraja, which claimed that the Khmer king was a representative of Shiva, the Hindu god of chaos, destruction, and rebirth. As such, in addition to the temples built to honor the Hindu gods, temples were also constructed to serve as tombs when kings died.

The Khmer built more than 100 stone temples spread out over about 40 miles. The temples were made from laterite (a material similar to clay that forms in tropical climates) and sandstone. The sandstone provided an open canvas for the statues and reliefs celebrating the Hindu gods that decorate the temples.

During the first half of the 12th century, Kambuja's King Suryavarman II decided to raise an enormous temple dedicated to the Hindu god Vishnu, a religious monument that would subdue the surrounding jungle and illustrate the power of the Khmer king. His masterpiece—the largest temple complex in the world—would be known to history by its Sanskrit name, "Angkor Wat," or "City of Temple."

Pilgrims visiting Angkor Wat in the 12th century would enter the temple complex by crossing a square, 600-foot-wide moat that ran some four miles in perimeter around the temple grounds. Approaching from the west, visitors would tread the moat's causeway to the main gateway. From there, they would follow a spiritual journey representing the path from the outside world through the Hindu universe and into Mount Meru, the home of the gods. They would pass a giant statue of an eight-armed Vishnu as they entered the western gopura, or gatehouse, known as the "Entrance of the Elephants." They would then follow a stone walkway decorated with nagas (mythical serpents) past sunken pools and column-studded buildings once believed to house sacred temple documents.

At the end of the stone walkway, a pilgrim would step up to a rectangular platform surrounded with galleries featuring six-foot-high bas-reliefs of gods and kings. One depicts the Churning of the Ocean of Milk, a Hindu story in which gods and demons churn a serpent in an ocean of milk to extract the elixir of life. Another illustrates the epic battle of monkey warriors against demons whose sovereign had kidnapped Sita, Rama's beautiful wife. Others depict the gruesome fates awaiting the wicked in the afterlife.

A visitor to King Suryavarman's kingdom would next ascend the dangerously steep steps to the temple's second level, an enclosed area boasting a courtyard decorated with hundreds of dancing apsaras, female images ornamented with jewelry and elaborately dressed hair.

For kings and high priests, the journey would continue with a climb up more steep steps to a 126-foot-high central temple, the pinnacle of Khmer society. Spreading out some 145 feet on each side, the square temple includes a courtyard cornered by four high conical towers shaped to look like lotus buds. The center of the temple is dominated by a fifth conical tower soaring 180 feet above the main causeway; inside it holds a golden statue of the Khmer patron, Vishnu, riding a half-man, half-bird creature in the image of King Suryavarman.

With the decline of the Khmer Empire and the resurgence of Buddhism, Angkor Wat was occupied by Buddhist monks, who claimed it as their own for many years. A cruciform gallery leading to the temple's second level was decorated with 1,000 Buddhas; the Vishnu statue in the central tower was replaced by an image of Buddha. The temple fell into various states of disrepair over the centuries. It's now the focus of international restoration efforts.

THE WEIRD WIDE WORLD

Q Can an animal be made out of glass?

A If you visit Harvard University's Museum of Natural History, you can view a small gallery that is filled with glass sea creatures that are so life-like, they look as if they were just pulled from an aquarium. Of course, they're not really alive. They are the work of Leopold and Rudolf Baschka, nineteenth-century artists who specialized in creating scientific models.

But if you drop by Dr. Joanna Aizenberg's Biomineralization and Biometrics Lab in the nearby Harvard School of Engineering and Applied Sciences, she will tell you that there are indeed real sea creatures that are made of glass. She has spent her career studying *euplectella aspergillum,* a sea sponge commonly known as Venus's flower-basket. This simple but intriguing organism weaves its lacy exoskeleton from naturally generated glass.

Silicon dioxide (a.k.a. silica), a chemical compound found in quartz and sand, is the main ingredient in glass that is made by humans. Seawater contains minute particles of silica, and the Venus's flower-basket siphons them into its cells and combines them with proteins to construct glass fibers that are two hundred nanometers wide, less than one-hundredth the thickness of a human hair. These tiny fibers are then glued together with additional enzymes to make rods for the flower-basket's lattice-like glass house. This smooth, hollow structure can rise nearly a foot from the ocean floor.

Unlike humans who live in glass houses, Venus's flower-basket has no need to fear stones or anything else that might be thrown its way. Its glass exoskeleton is extremely strong—a hundred times stronger than manmade glass, according to Aizenberg—and flexible. "You can bend them, twist them, and they probably won't break because the energy of the force you apply is dissipated in the glue," Aizenberg told MSNBC in 2005.

Most remarkably, the sponge makes glass without the use of heat. Human glassmaking can require temperatures of 3,100 degrees Fahrenheit or more. Venus's flower-basket can perform the same feat in tropical waters that are three hundred and fifty to a thousand feet deep, where temperatures might average only fifty degrees.

The precise process the flower-basket uses to engender glass is not clear. Professor Daniel Morse—director of the University of California, Santa Barbara, collaborative technologies program—believes that unraveling the mystery might give a boost to the fiber optics industry. Morse hopes to be able to grow fiberglass semiconductors in a test tube, in much the same way that Venus's baskets grow in the ocean. He is also examining other sponges that contain glass, such as the orange puffball, a small rock-clinging organism that keeps its glass fibers hidden inside its body.

Understanding the biochemistry of these sponges could also give a boost to the production of inexpensive, high-efficient solar-energy panels, Morse says.

It's mind-blowing to think that in the twenty-first century, people might be able to heat their homes using technology that was developed in the cold and lightless ocean millions of years ago.

Q What's the weirdest creature in the sea?

A Once you hit a certain depth, every sea creature is weird. There's the terrifying angler fish, famous for its appearance in the movie *Finding Nemo;* the purple jellyfish, lighting up the sea like a Chinese lantern; the horrid stonefish, with a face only a mother could love; and the straight-out-of-science-fiction chimaera, or ghost shark, with its long snout and venomous dorsal spine.

Yes, there are a lot of "weirdest creature" candidates down there. For the winner, we're going with one of the ocean's lesser-known oddities: the ominous vampire squid.

The sole member of the order Vampyromorphida, the vampire squid's scientific name is *Vampyroteuthis infernalis,* which translates literally into "vampire squid from Hell." The squid is as black as night and has a pair of bloodshot eyes. Full-grown, the squid is no more than a foot long. For its size, it has the largest eyes of any animal in the world. Its ruby peepers are as large as a wolf's eyes, sometimes more than an inch in diameter.

Like many deep-sea denizens, the vampire squid has bioluminescent photophores all over its body. The squid can apparently turn these lights on and off at will, and it uses this ability—combined with the blackness of its skin against the utter dark of the deep—to attract and disorient its prey.

The vampire squid is not a true squid—the order Vampyromorphida falls somewhere between the squid and the octopus—and so it does not possess an ink sac. In compensation, the vampire squid has the ability to expel a cloud of mucus when threatened; this mucus contains thousands of tiny bioluminescent orbs that serve to blind and confuse a predator while the vampire squid escapes into the shadows. As a second deterrent to predators, the vampire squid can turn itself inside out, exposing its suckers and cirri (tiny hair-like growths that act as tactile sensors) and making the creature look as though it is covered with spines.

Despite its name, the vampire squid does not feed on blood; its diet consists mostly of prawns and other tiny, floating creatures. Other than that, all that's missing for this Bela Lugosi mimic are the fangs and the widow's peak. But before you reach for a wooden stake, you should know that the vampire squid poses absolutely no threat to humans. It's found mostly at 1,500 to 2,500 feet below the surface, so the odds of encountering one are pretty slim.

Q Are there flesh-eating plants?

A Once upon a time, a hapless florist named Seymour Krelborn discovered a strange plant in his shop that had a taste for blood. If you're a fan of off-Broadway musicals, you'll recognize this as the plot of *Little Shop of Horrors.*

Seymour's potted pal might have been a fantasy, but flesh-eating plants do exist. The most famous is the Venus flytrap, whose pair of spiky, hinged petals snap shut on unsuspecting insects. Though the Venus flytrap might sound exotic, it's actually an all-American species that's found in the bogs and swamps of Florida, the Carolinas, and occasionally as far north as New Jersey.

Another well-known botanical carnivore is the pitcher plant. As its name implies, the plant's blossom is shaped like a narrow pitcher and filled with a deadly nectar that lures insects inside. Pitcher plants come in many varieties and are found in Europe, South America, and North America. One type, native to northern California and Oregon, has been dubbed the cobra plant, a sinister reference to its long, curved snake-like flower and mottled coloring.

The third-most common type of carnivorous plant is the so-called "flypaper" trap. The butterwort—found in Europe, North America, South America, Central America, and south Asia—is a good example of this variety. The thick, "buttery" leaves of this nondescript ground plant are coated with insect-trapping mucus.

Technically, all of these plants are insectivores, not carnivores. (*Carne* comes from the Latin word for "flesh.") So do any plants literally consume flesh? Large pitcher plants in South America have been known to digest frogs, small birds, and even tiny rodents, although this isn't their usual fare. People who cultivate carnivorous plants sometimes feed them bits of meat. But experts at the International Carnivorous Plant Society (ICPS) caution against this practice, pointing out that the enzymes these plants use to absorb nutrients are better adapted to insects and that the plants are likely to starve on a steady diet of beef and chicken.

Barry Rice, author of *Growing Carnivorous Plants,* confesses to feeding a few Venus flytraps fragments of skin that were sloughed from his own toes during a bout with athlete's foot. Much to his surprise, the traps took to these unappetizing tidbits. Unlike Audrey in *Little Shop of Horrors,* however, Rice's plants did not begin crooning, "Feed me! Feed me!" Be assured that any reports you hear of plants that are big enough to consume a human being are strictly figments of Hollywood's imagination.

In fact, carnivorous plants face a far greater threat from humans than vice versa. Development has threatened the habitats of many plants. Of the approximately 630 species of carnivorous plants identified by biologists, the ICPS estimates that twenty-six are currently imperiled. Carnivorous plants help balance the ecosystem by keeping insect populations in check. Without them, the planet might become a shop of horrors, indeed.

Q What are some of the world's strangest plants?

A When the first Western explorers returned from the Congo, they told tall tales of monstrous plants that demanded human flesh. Although we now know that no such plants exist, there are plenty of weird and scary plants in the world—enough for a little shop of horrors.

Kudzu: Native to China and Japan, when this vine was brought to the United States in 1876, its ability to grow a foot per day quickly made it a nuisance. With 400-pound roots, 4-inch-diameter stems, and a resistance to herbicides, it is nearly impossible to eliminate. Kudzu currently covers more than two million acres of land in the southern United States.

Cow's Udder: This shrub is known alternately as Nipple Fruit, Titty Fruit, and Apple of Sodom. (Did a group of 4th graders name it?) A relative of the tomato, it sports poisonous orange fruit that look like inflated udders.

King Monkey Cup: The largest of carnivorous pitcher plants traps its prey in pitchers up to 14 inches long and 6 inches wide. It then digests them in a half gallon of enzymatic fluid. The plant has been known to catch scorpions, mice, rats, and birds.

Titan Arum: Known in Indonesia as a "corpse flower," this plant blooms in captivity only once every three years. The six-foot-tall bloom weighs more than 140 pounds and looks, as its Latin name says, like a "giant shapeless penis." Even less appealingly, it secretes cadaverene and putrescine, odor compounds that are responsible for its smell of rotting flesh.

Resurrection fern: This epiphyte (that is, air plant) gets its nutrients and moisture from the air. Although other plants die if they lose 8 to 12 percent of their water content, the resurrection fern simply dries up and appears dead. In fact, it can survive despite losing 97 percent of its water content.

Wollemi pine: Previously known only through 90 million-year-old fossils, the Wollemi pine tree was rediscovered in Australia in 1994. Fewer than 100 adult trees exist today. Although propagated trees are being sold around the world, the original grove's location is a well-guarded secret, disclosed to only a few researchers.

Rafflesia arnoldii: Also known as a "meat flower," this parasitic plant has the largest single bloom of any plant, measuring three feet across. It can hold several gallons of nectar, and its smell has been compared to "buffalo carcass in an advanced stage of decomposition."

Hydnora africana: This parasitic plant is found in Namibia and South Africa growing on the roots of the Euphorbia succulent. Most of the plant is underground, but the upper part of the flower looks like a gaping, fang-filled mouth. And, because smelling like rotting flesh is de rigueur in the weird-plant world, it emits a putrid scent to attract dung or carrion beetles.

Aquatic duckweed: Also known as watermeal because it resembles cornmeal floating on the surface of water, this is the smallest flowering plant on earth. The plant is only .61 millimeter long, and the edible fruit, similar to a (very tiny) fig, is about the size of a grain of salt.

Spanish moss: Although it is a necessary prop in Southern Gothic horror tales, Spanish moss is neither moss nor Spanish. Also known as Florida moss, long moss, or graybeard, it is an air plant (epiphyte) that takes nutrients from the air. It's also related to the pineapple and has been used to stuff furniture, car seats, and mattresses.

Baobab tree: The baobab is the world's largest succulent, reaching heights of 75 feet. It can also live for several thousand years. The tree's strange, root-like branches gave rise to the legend that they grow upside down. Their enormous trunks are often hollowed out and used as shelter, including a storage barn, an Australian prison, a South African pub, bus stops, and in Zambia, a public toilet (with flushing water, no less).

Q Can you outrun lava?

A It depends on how fast you can run and how fast the lava flows. The absolute fastest humans in the world can run a little faster than ten meters per second, but only for one hundred meters. For a five-thousand-meter Olympic race, peak human performance is just more than six meters per second.

Assuming you'll have a major adrenaline boost due to the dire circumstances, we'll say that you can maintain a speed of three to five meters per second. This speed could vary greatly, however, depending on your physical condition and the distance you need to run, which might be several kilometers.

The speed of lava is affected by its temperature and viscosity (which are related), the angle of the slope it is flowing, and the expulsion rate of the volcano.

There are different types of volcanoes and varieties of lava. Some, you could probably outwalk; other types of lava would swallow and incinerate an Olympic-class runner before he or she took a single step.

A pyroclastic flow isn't actually made of lava—it's a column of hot ash and gas that collapses under its own weight and roars down the side of the volcano like an avalanche. These flows can reach speeds of forty meters per second—you have no chance of outrunning them.

Basaltic lava has a high temperature and low viscosity, which means that it can move quickly, approaching speeds of thirty meters per second. However, many basaltic flows are much slower—two meters per second or less. You could outpace it for a while, but basaltic lava is relentless and often flows ten or more kilometers from the volcano before cooling and coming to a stop. You might outrun the slower flows, but it would be a challenge.

Mount Kilauea in Hawaii has been continuously issuing basaltic lava flows since 1983. Occasionally, the flows extend to nearby towns, most of which have been abandoned. When there are Hawaiians in the path of the lava, however, they are able to run away from the generally slow flows.

Rhyolitic lava moves very slowly because it has a relatively low temperature and high viscosity. It may move only a few meters in an hour. It is still dangerously hot, however, so while you can easily outpace it with a brisk walk, you shouldn't dilly-dally.

Q What is the coldest temperature possible?

A Put it this way: The little thermometer outside your kitchen window that you picked up at Wal-Mart couldn't begin to display the coldest temperature possible. But if it could, it would read -273.15 degrees Celsius (or -459.67 degrees Fahrenheit, for those who prefer that temperature scale). That's as cold as it could ever get anywhere in the universe. It's when molecular movement stops, which is how scientists define the lowest temperature possible.

On the Kelvin scale, this temperature is called, fittingly, absolute zero. What is the Kelvin scale? It is named after Lord Kelvin, an Irish mathematician and physicist who created the scale in the mid-eighteen hundreds because he felt the world needed a definitive way to measure "infinite cold." It goes up and down in the same increments as the Celsius scale—the zero point is simply different. On the Celsius scale, zero is the temperature at which water freezes. (Freezing water on the Kelvin scale is 273.15 K.)

It's impossible to chill something to absolute zero artificially, but that hasn't stopped the men and women in white lab coats from trying. Funny things happen at temperatures approaching zero K, including superfluidity, in which a liquid such as helium loses all friction and moves in and around containers seemingly against the laws of gravity; superconductivity, in which a substance loses all resistance to electrical impulse; and Bose-Einstein condensate, in which even subatomic movement slows.

The coldest place scientists have found in the universe is the Boomerang Nebula. It's an ultrafrigid one degree Kelvin (or -272 degrees Celsius). Fortunately, no one from Earth will be going there anytime soon. It's in the constellation Centaurus, about five thousand light years away.

Q Can you be killed by a plant?

A The good news is, unless you're a character in *Little Shop of Horrors,* no plant is going to kill you with malicious intent or for food. That doesn't mean you shouldn't fear death by plant. Obviously, a tree could fall on you or twist your car into a pretzel if you veer off the road. But the more gruesome scenarios involve eating something you shouldn't.

Here's a sampling from the menu of green meanies:

• Aconitum (aconite, monkshood, or wolfsbane) will start your mouth burning from the first nibble. Then you'll start vomiting, your lungs and heart will shut down, and you'll die of asphyxiation. As luck would have it, your mind will stay alert the entire time. And you don't even have to eat aconitum to enjoy its effects: Just brush up against it and the sap can get through your skin.

• Hemlock is another particularly nasty snack. In fact, the ancient Greeks gave it to prisoners who were condemned to die (including Socrates). Ingest some hemlock and it will eventually paralyze your nervous system, causing you to die from lack of oxygen to the brain and heart. Fortunately, if you happen to have an artificial ventilation system nearby, you can hook yourself up and wait about three days for the effects to wear off. But even if there is a ventilation system handy, it's best if you just don't eat hemlock.

• Oleander is chock full of poisony goodness, too. Every part of these lovely ornamental plants is deadly if ingested. Just one leaf can be fatal to a small child, while adults might get to enjoy up to ten leaves before venturing into the big sleep. Even its fumes are toxic—never use Oleander branches as firewood. Oleander poisoning will affect most parts of your body: the central nervous system, the skin, the heart, and the brain. After the seizures and the tremors, you may welcome the sweet relief of the coma that might come next. Unfortunately, that can be followed by death.

So, while there is no need to worry about any plants sneaking up on you from behind with a baseball bat, there are plenty of reasons not to take a nibble out of every plant you see.

Q What is the slowest-moving object in the world?

A Jet cars and supersonic airplanes get all the glory for their high-speed records, but there are some objects that are just as notable for their amazing slowness. In fact, they go so slowly that scientists need special equipment to detect their movement. What moves slowest of all? The answer just might be right under your feet.

The surface of the earth is covered by tectonic plates, rigid slabs made of the planet's crust and the brittle uppermost mantle below, called the lithosphere. Some of the plates are enormous, and each is in constant movement—shifting, sliding, or colliding with other plates or sliding underneath to be drawn back down into the deep mantle. The plates "float" on the lower mantle, or asthenosphere; the lower mantle is not a liquid, but it is subjected to heat and pressure, which softens it so that it can flow very, very slowly.

When an earthquake occurs, parts of the plates can move very suddenly. Following the Great Alaska Earthquake in 1964, America's largest ever, the two plates involved shifted about thirty feet by the end of the event. However, most of the time tectonic plates move relatively steadily and very slowly. Scientists use a technique called Satellite Laser Ranging (SLR) to detect their movement.

SLR relies on a group of stations spread around the world that use lasers to send extremely short pulses of light to satellites equipped with special reflective surfaces. The time it takes for the light to make the roundtrip from the satellite's main reflector is instantaneously measured. According to the U.S. Geological Survey, this collection of measurements "provides instantaneous range measurements of millimeter level precision" that can be used in numerous scientific applications. One of those applications is measuring the movement of the earth's tectonic plates over time.

How slow do tectonic plates move? The exact speed varies: The slowest plates move at about the same rate of speed that your fingernails grow, and the fastest plates go at about the same rate that your hair grows. A rough range is one to thirteen centimeters per year. The fastest plates are the oceanic plates, and the slowest are the continental plates. At the moment, the Slowest Object Award is a tie between the Indian and Arabian plates, which are moving only three millimeters per year.

If you're wondering who the runner-up is in the race to be slowest, it appears to be glaciers. The slowest glaciers creep a few inches each day, still faster than tectonic plates. However, some glaciers are so speedy they can cover nearly eight miles in a single year, and sometimes a glacier can surge. In 1936, the Black Rapids Glacier in Alaska galloped toward a nearby lodge and highway, averaging fifty-three meters a day over three months. That leaves tectonic plates in the dust.

Q What are some of the stranger structures in North America?

A Humans have the capacity to achieve great things, conquer the seemingly impossible, and invent wonders that make our world a better place. However, sometimes they just like to build things that are big, tall, or strange.

Pratt Rocks: South Dakota has Mount Rushmore, but nestled in the Catskill Mountains is the town of Prattsville, New York, which features Pratt Rocks—a set of relief carvings begun 84 years before its famous western counterpart. Zadock Pratt, who founded the world's largest tannery in the 1830s, commissioned a local sculptor to immortalize his visage high up on a mountainside. The numerous stone carvings include a coat of arms, Pratt's own bust, his business milestones, and even his personal accomplishments, such as his two terms in the U.S. House of Representatives. Carvings also include a shrine to Pratt's son George, who was killed during the Civil War. But the strangest bit found at this site is a recessed tomb that was intended to house Pratt's decaying corpse for eternity. It leaked, Pratt balked, and the chamber remains empty.

Columcille: The offbeat dream of Bill Cohea, Jr., and Frederick Lindkvist, two highly spiritual fellows, Columcille was designed to resemble an ancient Scottish religious retreat located on the Isle of Iona. More than 80 oblong stones are "planted" in a Bangor, Pennsylvania, field to approximate the ancient site, a place where some say "the veil is thin between worlds." In addition to the megaliths, Columcille has enchanting chapels, altars, bell towers, cairns, and gates—enough features to lure Harry Potter fans into an entire day of exploration. Cohea and Lindkvist began their ever-evolving project as a spiritual retreat in 1978. They encourage everyone to visit their nondenominational mystical park. Their request? Simply "be."

Homer City Generating Station: It's quite surprising to encounter a 1,216-foot-tall smokestack, especially when that chimney is located in a rural town deep in western Pennsylvania. Homer City Generating Station produces electricity by burning coal. But the process has one troubling side effect: Its effluence can be toxic in certain quantities. The super-tall smokestack's purpose is to harmlessly disperse this undesirable by-product, thereby rendering it safe. It does this by releasing the agents high up in the atmosphere where they (theoretically) have ample time to dilute before falling back to Earth. At present, this soaring chunk of steel-reinforced concrete ranks as the third tallest in the world, just behind a 1,250-foot-tall smokestack located in Canada and a 1,377-foot-tall monster over in Kazakhstan.

Staunton, Virginia: A drive through Staunton, Virginia, may leave some wondering if they've mistakenly entered the land of the giants. After all, an 18-foot-tall watering can and a six-foot-tall flowerpot are displayed on the main boulevard. But fear not. It's no giant who dwells in this hamlet but rather an average-size gent named Willie Ferguson. A large concentration of this metal fabricator's giant works can be seen on the grounds of his sculpture studio. At this metal "imaginarium," visitors will find a six-foot-long dagger, a ten-foot-long set of crutches, a six-foot-tall work boot—everything, it seems, but the proverbial beanstalk.

A Floating Bridge: If you've crossed Vermont's Brookfield Floating Bridge by car, you're aware of its treachery. If you tried it on a motorcycle, you probably took an unplanned swim. That's because the lake that the bridge is supposed to cross occasionally crosses it. The 320-foot-long all-wooden Brookfield Bridge rests on 380 tarred, oaken barrels that were designed to adjust to the level of Sunset Lake and keep the bridge deck high and dry. But more often than not, they allow the bridge to sink several inches below the surface. Why does this bridge float in the first place? Sunset Lake is too deep to support a traditional, pillared span, so since 1820, impromptu "water ballet" maneuvers have been taking place as vehicles amble across.

Roxborough Antenna Farm: As drivers creep along I-76 just west of Philadelphia, they witness a stand of super-tall broadcasting masts towering over a suburban neighborhood. The Roxborough Antenna Farm is to broadcasting towers what New York City is to skyscrapers. In the land of broadcasting, height equals might, so the higher the tower, the better the signal strength. With eight TV/FM masts jutting above the 1,000-foot mark (the tallest stretches to 1,276 feet), the array easily outclasses most skyscrapers in height. The reason these big sticks exist in such a concentrated area? Location, location, location. The Roxborough site is a unique setting that features geographical height, proper zoning clearances, and favorable proximity to the city—a trifecta by industry standards.

Can beer flood?

A Life is full of surprises, some less pleasant, some more so—including, yes, a flood of beer. From molasses rivers to raining frogs to exploding whales, headlines continually prove that truth is sometimes stranger than fiction. Let's look at some strange catastrophes from around the world.

The London Beer Flood: In 1814, a vat of beer erupted in a London brewery. Within minutes, the explosion had split open several other vats, and more than 320,000 gallons of beer flooded the streets of a nearby slum. People rushed to save as much of the beer as they could, collecting it in pots, cans, and cups. Others scooped the beer up in their hands and drank it as quickly as they could. Nine people died in the flood—eight from drowning and one from alcohol poisoning.

The Great Siberian Explosion: Around 7:00 A.M. on June 30, 1908, 60 million trees in remote Siberia were flattened by a mysterious 15-megaton explosion. The huge blast, which occurred about five miles above the surface of the earth, traveled around the world twice and triggered a strong, four-hour magnetic storm. Magnetic storms occur about once every hundred years, and can create radiation similar to a nuclear explosion. These storms start in space and are typically accompanied by solar flares.

The 1908 explosion may have started with a comet of ice, which melted and exploded as it entered Earth's atmosphere. Or, it may have been an unusual airburst from an asteroid. Others believe that the source was a nuclear-powered spacecraft from another planet. However, no physical evidence of the cause has ever been found.

The Boston Molasses Disaster: On an unusually warm January day in 1919, a molasses tank burst near downtown Boston, sending more than two million gallons of the sticky sweetener flowing through the city's North End at an estimated 35 miles per hour. The force of the molasses wave was so intense that it lifted a train off its tracks and crushed several buildings in its path. When the flood finally came to a halt, molasses was two to three feet deep in the streets, 21 people and several horses had died, and more than 150 people were injured. Nearly 90 years later, people in Boston can still smell molasses during sultry summer days.

It's Raining…Frogs: On September 7, 1953, clouds formed over Leicester, Massachusetts—a peaceful little town near the middle of the state. Within a few hours, a downpour began, but it wasn't rain falling from the sky—thousands of frogs and toads dropped out of the air. Children collected them in buckets as if it was a game. Town officials insisted that the creatures had simply escaped from a nearby pond, but many of them landed on roofs and in gutters, which seemed to dispute this theory. It is still unclear why the frogs appeared in Leicester or why the same thing happened almost 20 years later in Brignoles, France.

Oregon's Exploding Whale: When an eight-ton sperm whale beaches itself in your town, what do you do? That's a question residents of Florence, Oregon, faced in November 1970. After consulting with the U.S. Navy, town officials decided to blow up the carcass with a half ton of dynamite. Spectators and news crews gathered to watch but were horrified when they were engulfed in a sandy, reddish mist and slapped by flying pieces of whale blubber. A quarter mile away, a car was crushed when a gigantic chunk of whale flesh landed on it. No one was seriously hurt in the incident, but when the air cleared, most of the whale was still on the beach. The highway department hauled the rest of it away.

Q What are some of the noisiest animals?

A Animals send out messages for very specific reasons, such as to signal danger or for mating rituals. Some of these calls, like the ones that follow, are so loud they can travel through water or bounce off trees for miles to get to their recipient.

Blue Whale: The call of the mighty blue whale is the loudest on Earth, registering a whopping 188 decibels. (The average rock concert only reaches about 100 decibels.) Male blue whales use their deafening, rumbling call to attract mates from hundreds of miles away.

Howler Monkey: Found in the rain forests of the Americas, this monkey grows to about four feet tall and has a howl that can travel more than two miles.

Elephant: When an elephant stomps its feet, the vibrations created can travel 20 miles through the ground. They receive messages through their feet, too. Research on elephants has identified a message for warning, another for greeting, and another for announcing, "Let's go." These sounds register from 80 to 90 decibels, which is louder than most humans can yell.

North American Bullfrog: The name comes from the loud, deep bellow that male frogs emit. This call can be heard up to a half mile away, making them seem bigger and more ferocious than they really are. To create this resonating sound used for his mating call, the male frog pumps air back and forth between his lungs and mouth, and across his vocal cords.

Hyena: If you happen to hear the call of a "laughing" or spotted hyena, we recommend you leave the building. Hyenas make the staccato, high-pitched series of hee-hee-hee sounds (called "giggles" by zoologists) when they're being threatened, chased, or attacked. This disturbing "laugh" can be heard up to eight miles away.

African Lion: Perhaps the most recognizable animal call, the roar of a lion is used by males to chase off rivals and exhibit dominance. Female lions roar to protect their cubs and attract the attention of males. Lions have reportedly been heard roaring a whopping five miles away.

Northern Elephant Seal Bull: Along the coastline of California live strange-looking elephant seals, with huge snouts and big, floppy bodies. When it's time to mate, the males, or "bulls," let out a call similar to an elephant's trumpet. This call, which can be heard for several miles, lets other males—and all the females nearby—know who's in control of the area.

Q What's the most dangerous animal on Earth?

A Though some argue that humans are the most dangerous creatures on Earth, the distinction actually belongs to a tiny insect. Diseases transmitted via mosquito bites have caused more death and misery than the total number of casualties and deaths suffered in all of history's wars.

The Roman Empire crumbled in the 3rd and 4th centuries when the Legions, decimated by malaria, were unable to repel barbarian invaders. American planners took control of the Panama Canal construction project when French interests withdrew after losing more than 22,000 workers to mosquito-borne illnesses over an eight-year period.

More than 2,500 species of mosquito spread disease throughout the world. The killers begin life as larvae, hatched in almost any kind of water. Within one week, adults emerge to ply their deadly trade.

As with other species, the female is the deadlier of the two sexes, while the male concerns himself with nothing more than fertilizing eggs. Females need blood to nourish their eggs; the process of collecting this food supply can wreak havoc on humans.

Mosquitoes can fly more than 20 miles from the water source in which they were born. Sensory glands allow the insects to detect carbon monoxide exhaled by their victims and lactic acid found in perspiration. When a female mosquito dips her proboscis into an unwilling victim, she transfers microorganisms through her saliva into her donor. These are responsible for some of the world's most deadly and debilitating diseases, which include malaria; Yellow, Dengue, and Rift Valley fevers; West Nile virus; and at least six different forms of encephalitis.

According to the U.S. Department of Health and Human Sciences, more than 500 million cases of malaria are reported worldwide each year, resulting in an average of one million deaths.

Q Were the ancient Egyptians the first civilization to practice mummification?

A Turns out that the Egyptians—history's most famous embalmers—weren't the first. By the time Egyptians were fumbling with the art, Saharans and Andeans were veterans at mortuary science.

In northern Chile and southern Peru, modern researchers have found hundreds of pre-Inca mummies (roughly 5000–2000 B.C.) from the Chinchorro culture. Evidently, the Chinchorros mummified all walks of life: rich, poor, elderly, didn't matter. We still don't know exactly why, but a simple, plausible explanation is that they wanted to honor and respect their dead.

The work shows the evolution of increasingly sophisticated, artistic techniques that weren't very different from later African methods: Take out the wet stuff before it gets too gross, pack the body carefully, dry it out. The process occurred near the open-air baking oven we call the Atacama Desert, which may hold a clue in itself.

The oldest known instance of deliberate mummification in Africa comes from ancient Saharan cattle ranchers. In southern Libya, at a rock shelter now called Uan Muhuggiag, archaeologists found evidence of basic seminomadic civilization, including animal domestication, pottery, and ceremonial burial.

We don't know why the people of Uan Muhuggiag mummified a young boy, but they did a good job. Dispute exists about dating here: Some date the remains back to the 7400s B.C., others to only 3400s B.C. Even at the latest reasonable dating, this predates large-scale Egyptian practices. The remains demonstrate refinement and specialized knowledge that likely took centuries to develop. Quite possibly some of this knowledge filtered into Egyptian understanding given that some of the other cultural finds at Uan Muhuggiag look pre-Egyptian as well.

Some 7,000 to 12,000 years ago, Egyptians buried their dead in hot sand without wrapping. Given Egypt's naturally arid climate, the corpse sometimes dehydrated so quickly that decay was minimal. Sands shift, of course, which would sometimes lead to passersby finding an exposed body in surprisingly good shape. Perhaps this inspired early Egyptian mummification efforts.

As Egyptian civilization advanced, mummification interwove with their view of the afterlife. Professionals formalized and refined the process. A whole industry arose, offering funerary options from deluxe (special spices, carved wood case) to budget (dry 'em out and hand 'em back). Natron, a mixture of sodium salts abundant along the Nile, made a big difference. If you extracted the guts and brains from a corpse, then dried it out it in natron for a couple of months, the remains would keep for a long time. The earliest known Egyptian mummy dates to around 3300 B.C.

It's hard to ignore a common factor among these cultures: proximity to deserts. It seems likely that ancient civilizations got the idea from seeing natural mummies.

Ice and bogs can also preserve a body by accident, of course, but they don't necessarily mummify it. Once exposed, the preservation of the remains depends on swift discovery and professional handling. If ancient Africans and South Americans developed mummification based on desert-dried bodies, it would explain why bogs and glaciers didn't lead to similar mortuary science. The ancients had no convenient way to deliberately keep a body frozen year-round without losing track of it, nor could they create a controlled mini-bog environment. But people could and did replicate the desert's action on human remains.

We make mummies today, believe it or not. An embalmed corpse is a mummy—it's just a question of how far the embalmers went in their preservation efforts. To put it indelicately: If you've attended an open-casket funeral, you've seen a mummy.

OUR WORLD'S WATERWAYS

Q Is the Red Sea really red?

A Standing on the shore of the Red Sea, you might wonder how such intensely blue-green waters could be so obviously mislabeled with the moniker "Red." Is it possible the person who named this 1,200-mile strip of sea, located between Africa and Asia, suffered from an acute case of colorblindness? No, it's more likely he or she saw the Red Sea while the *Trichodesmium erythraeum* was in full bloom. Before you get all excited, *Trichodesmium erythraeum* is not some kind of wildly exotic orchid indigenous to Egypt. It's simply a type of cyanobacteria, a.k.a. marine algae.

You've seen how an overgrowth of algae can turn your favorite pond or motel pool a murky shade of opaque green, right? In the case of the Red Sea, the alga is rich in a red-colored protein called phycoerythrin. During the occasional bloom, groups of red- and pink-hued *Trichodesmium erythraeum* blanket the surface of the sea. When they die off, they appear to transform the waters from a heavenly shade of blue to a rustier reddish-brown.

While this algae-induced color change is a widely accepted derivation for the Red Sea's name, there is another theory: Some say mariners of antiquity were inspired by the region's mineral-rich red mountain ranges and coral reefs, so they named the body of water *Mare Rostrum* (Latin for "Red Sea"). In 1923, English author E. M. Forster agreed, describing the Red Sea as an "exquisite corridor of tinted mountains and radiant water."

However you choose to color it, one fact still remains: Beneath a ruddy exterior, there lies a deep Red Sea that is actually true blue.

Q What's the difference between a lake and a pond?

A Lakes and ponds are both bodies of water that fill depressions in the earth's surface. But what's the difference between, say, the majestic Lake Champlain and that sinkhole over yonder?

According to the U.S. Environmental Protection Agency, the main distinction is size. You could probably walk all the way across a pond without getting submerged—especially if it's that ditch you dug up and filled with goldfish in your backyard. Often, ponds are human-made.

Lakes, on the other hand, occur naturally. And to cross one, you'll need a boat or a really mean backstroke. In some cases, you can't even see across to the other side. Just bring your landlocked cousin up for a visit to see the great Lake Huron. Chances are, he or she will say, "That's no lake. That's the ocean!"

Because of their size, lakes are too deep to support any rooted plants, except maybe near the shore. Sunlight just can't penetrate to the bottom of Mother Nature's natatoriums. Lake Superior, for example, reaches a maximum depth of about 1,333 feet. Shallower ponds, however, usually have pondweed and vegetation growing along the bottom and around the edges.

Not only is the pond floor nice and muddy, there's little wave action to disrupt rooted plants or those bladderworts that float freely on the surface.

When it comes to the actual water, ponds tend to be the same temperature from top to bottom. And because they're small and shallow, ponds are greatly influenced by the local weather. In cold winter months, an entire pond can freeze solid. For sports enthusiasts in "tundric" territories, that means it's time to get out the ice skates and hockey sticks.

On the other hand, large lakes can have an impact on the local climate. People living around the Great Lakes region of the United States and Canada are all too familiar with weather forecasts of "cooler by the lake" and "lake-effect snow." But even in cold climates, most lakes are too large to freeze solid. Lace up your blades on Lake Michigan, and you might be skating on very thin ice!

Q When did piracy begin?

A Probably 15 minutes after the first ancient meeting of two river canoes. From the European perspective, the Golden Age of Piracy began around A.D. 1660; by 1730, the problem was largely under control. Here are some of the most notorious rascals of this era, plus noteworthy pirates from other times.

John Taylor: English. Taylor earned his fame capturing the Portuguese carrack Nossa Senhora do Cabo in 1721. It was one of the richest prizes of its time, consisting of diamonds and other portable loot. No idiot, Taylor then bought a pardon in Panama, where he likely retired wealthy.

"Blackbeard" Edward Teach: English. The infamous Blackbeard was a full-time drunkard who terrorized the Carolina coast, racking up captures and loot. But it wasn't all about money and rum: He once raided Charleston and took hostages until townsfolk gave him much-needed medicine for his venereal disease. In 1718, the Royal Navy ran down and killed Blackbeard just off Ocracoke Island.

Thomas Tew: English. His 1692–1694 Indian Ocean rampage made Tew rich enough to retire in New York under the protection of his friend, Governor Benjamin Fletcher. In 1695, his former crew talked him into one more voyage to raid Moghul ships. Piracy was a little like gambling: Most people lose, and winners should probably quit. Tew didn't, and in 1695 he was disemboweled by cannon fire.

Jean-David Nau: French. Also known as François l'Olonnais, Nau was a psycho in a sick line of work. Nau whittled prisoners with his cutlass, once allegedly eating a beating heart. Let us all thank the Native Central Americans for tearing him apart in 1668.

"Calico" Jack Rackham, Anne Bonny, and Mary Read: English, Irish, and English. The fancy-dressing Rackham got his buccaneering start under the notorious Charles Vane, whom he soon deposed in a mutiny. Rackham fell in love with Anne Bonny, who was stuck in an unhappy marriage in the Bahamas. They eloped and began a new piratical career aboard the sloop Revenge, where Bonny's hard-living, hard-fighting style earned her respect in a male-dominated business.

Anne soon discovered another incognita in men's clothing: Mary Read, who became her close friend—and her peer in the fine arts of fighting, swearing, and drinking. When pirate hunter Jonathan Barnet caught up with Revenge in 1720, Read and Bonny were among the few crewmembers sober and/or brave enough to fight.

As Rackham was marched off to hang, Bonny showed her contempt: "If ye'd fought like a man, ye needn't hang like a dog." (One suspects that history has omitted a volley or two of choice profanities.)

Both women escaped Calico Jack's fate by revealing their pregnancies to the court. In 1721, Read died amid disgusting jail conditions, but Bonny was most likely ransomed by her father.

Charles Vane: English. When the reformer Woodes Rogers showed up to clean out the Caribbean pirates' favorite lair (New Providence, now Nassau), the wily Vane was the only captain to reject the proffered amnesty. He kept marauding—losing ships and gaining them—until his luck and cunning finally ran out. He was captured and hanged in 1720.

Edward Low: English. Low was hideously scarred by a cutlass slash to the jaw, but the tortures he inflicted upon others would have nauseated a Stalin-era KGB interrogator. It's said that he once forced a man to eat his own severed ears—with salt. According to some sources, he was last seen running from a Royal Navy warship; good thinking on his part, given his record. It is believed he was hanged by the French after 1723.

George Lowther: English. Lowther was elected captain after a mutiny aboard a slave ship. This pirate's hobbies were torture and rape. In 1723, while his ship was beached for hull maintenance, the Royal Navy showed up. Lowther decided he'd rather not dangle from the gallows and shot himself instead.

Sir Henry Morgan: Welsh. Spain was the dominant colonial power in the Americas, shipping home gold and silver by the boatload. Morgan served English foreign interests by raising merry hell with Spanish shipping and colonial interests. Moral of his story: If your piracy serves the national interest and makes you rich, your country may reward you with a knighthood and a governorship. If you're Morgan, you then gain lots of weight and drink yourself to death as a pillar of the genteel Jamaican community. He died a wealthy man, succumbing to dropsy or liver failure in 1688.

Olivier "la Buze" Levasseur: French. "The Buzzard" collaborated with John Taylor and actually wore an eye patch. Unlike Taylor, though, Levasseur didn't quit while he was ahead but kept up his piratical career until his capture by French authorities at Madagascar in 1730. He was hanged.

Gráinne ni Mháille: Irish, aka Grace O'Malley. Rebel, seagoing racketeer, admiral—she was all these and more, engaging in piracy to champion the Irish cause against England. Gráinne didn't even kowtow to Queen Elizabeth I, though she did visit for tea and chitchat. She died of old age in 1603.

Howell Davis: Welsh. This sneaky rogue once captured two ships in one encounter. After the first catch, he forced the captives to brandish weapons at the second, inflating his apparent numbers. Davis was planning to seize a Portuguese island governor when the local militia recognized him and shot him in an ambush in 1719.

Rachel (Schmidt) Wall: American. She turned pirate with her husband, George, luring likely prizes with convincing distress cries. After George drowned in a storm, Rachel forsook the sea for petty thieving along the Boston docks. In 1789, she was arrested for trying to steal a woman's bonnet and was then accused of murdering a sailor. She stood trial, confessed to piracy, and was hanged.

Zheng Yi Sao: Chinese, aka Ching Shih. Hundreds of pirate ships sailed the Chinese coast under her command. Rare in piracy, she enforced strict rules—notably, a prohibition on rape. Her fleet dominated the coasts to such a degree that in some places, it functioned as a government. Zheng quit while she was ahead, swapping her fleet for a pardon. She died of natural causes in 1844.

Q Why do pirates wear eye patches?

A Isn't it obvious? It's one more place to put a jewel or gem. Nothing keeps eye contact like the beauty and shimmer of a diamond or ruby, and pirates were all about accessories.

Okay, the real answer is that most pirates didn't have eye patches. Read the biographies of most famous pirates in history—Blackbeard, Bartholomew Roberts, Calico Jack, and others—and you will find that their portraits show them with two patch-free eyes. So the main reason that pirates wear eye patches is because the creators of fictional pirates like them that way. Patches have become an instantly recognizable part of pirate lore, largely because pirates are portrayed as wearing them in many movies and books.

But that's not to say no pirate ever wore an eye patch. The primary reason anyone wears an eye patch—because he or she has a missing or injured eye—is a plausible explanation for why a pirate would wear one. Swashbuckling, after all, is an extremely dangerous activity.

Another, more fascinating theory involves a "trick of the trade" for all seamen, as well as for law-enforcement officers and armed forces. An eye patch (or simply closing one eye, which doesn't look nearly as cool) can help your eyesight when moving from a bright place into a dark one. For example, say you're a pirate and you've just boarded a ship at noon on a beautiful sunny day. If you're not wearing an eye patch and everyone you need to steal from is below deck, you could be compromised because your eyes have to become accustomed to the dark interior.

Not so with an eye patch! A patch gives you one eye that is already used to the dark. So when you get down below, you simply move the patch from one eye to the other. This allows for much more efficient pillaging.

In fact, you can use this trick the next time you go to see a pirate-themed movie: Close one eye while you're walking through the lobby, and then open it after you are in the theater. Call yourself a pirate, and find yourself a seat in the dark.

Q Why is the pirate flag called a Jolly Roger?

A With the 1883 publication of Robert Louis Stevenson's *Treasure Island,* the popular idea of the pirate germinated: a witty rogue with an eye patch, a peg-leg, and a smart-ass parrot, sailing the seven seas under the Jolly Roger, good-naturedly plundering booty and instigating a little plank-walking. Unfortunately for those romantics who long for the swashbuckling days of yore, most of Bob Lou Steve's details aren't particularly accurate.

There is little evidence that something as dramatic as "walking the plank" happened much, and parrots were rarely recorded as ships' mascots. But calling the pirate flag the "Jolly Roger" was one of the details Stevenson got right.

For hundreds of years, ships have hoisted the colors of their home country to let other ships know from where they hail. In the golden age of piracy, pirates used this form of communication as well, though more deviously. Often, pirates would fly flags of certain countries as a form of deception, in order to get close to their prey. Once they were within striking distance, the buccaneers would lower their false flags and raise their own ensigns. These flags varied from pirate to pirate, but they all meant the same thing: "Surrender, hand over your booty, and we will not kill you." Though if the pirates raised a red flag, it meant, "We will kill you and take your booty." (One might say these flags were the original "booty call.")

French pirates most prominently used the red flag as a symbol of imminent death, and among these pirates, such a flag became known as a *joli rouge* ("pretty red"). The English, hewing to their long tradition of making no effort to correctly pronounce foreign words, turned this into the "Jolly Roger."

Another theory points to a legendary Tamil pirate by the name of Ali Raja. Raja ruled the Indian Ocean and had such a reputation that even English seamen had heard of the pirate captain. It's not hard to imagine how Europeans who were unfamiliar with Middle Eastern languages might corrupt "Ali Raja" into "Jolly Roger."

The least interesting hypothesis points to the fact that in England during piracy's glory days, the devil was often referred to as "Old Roger." That, combined with the grinning appearance of the skull symbol, led to the flag being called the "Jolly Roger." Unfortunately, there is no definitive evidence that supports one theory over another.

The origin of the familiar skull-and-crossbones image is also unclear. The image had been used as a general symbol of death long before pirates appropriated it—crusaders used the symbol in the 1100s, for example. The first recorded use of the skull-and-crossbones on a pirate flag was in 1700, when a French buccaneer named Emmanuel Wynne hoisted it. After that, the black flag with a variation of the image appeared more frequently and sometimes included hourglasses, spears, and dancing skeletons.

Once Stevenson published *Treasure Island,* the skull-and-crossbones—along with the mythical parrot—became forever associated with pirates in the popular mind. The novel is also famous for introducing the phrase, "Yo, ho, ho, and a bottle of rum" into pirate lore. We don't know what that means, either.

Q Why doesn't water in a water tower freeze?

A Although our chemistry teacher more likely compared us to Linus Van Pelt than Linus Pauling, even we know that water freezes at thirty-two degrees Fahrenheit. Which is why we've never been able to understand why water doesn't freeze in water towers during those long, cold winters.

To fully understand why you're able to turn on your faucet and have running water even on the coldest days, we need to look at how water towers work. Most towns get their water from wells or bodies of water such as lakes. This water is pumped to a water treatment plant, where it is disinfected before being delivered through a main pipeline to the rest of the area's delivery system.

A water tower is hooked up to that system, drawing water into its reservoir as it is pumped through the main pipes. When the demand for water is too much for the system pump to handle, gravity and water pressure release water from the tower back into the main pipeline. During off-peak times, the water tower refills from the pipeline.

This is a simple, efficient system, and one that helps explain why water towers don't freeze solid in the winter. When a water tower pulls water from the pipeline to refill its reservoir, it is drawing somewhat warmer water from the pipes. Furthermore, water towers are drained and refilled fairly frequently, making it difficult for ice to form. The agitation of water molecules from the movement of draining and refilling slows down the freezing process, too. (To get an idea of the way this works, think of how long it takes waterfalls or rivers to freeze.)

However, in some parts of the country—such as the frozen tundra of North Dakota—water in water towers does freeze. Rarely, though, does it freeze solid. In climates where freezing is a danger, water towers are more heavily insulated, and some are even built with heating systems near their bases that prevent water from freezing on its way into the tower.

Of course, no precautions are foolproof. Just about anything can freeze over if it's cold enough for long enough: lakes, waterfalls, water towers, and, if our passing grade in high school chemistry is any indication, even hell.

Q Do rivers always flow north to south?

A No, rivers are not subject to any natural laws that compel them to flow north to south. Only one thing governs the direction of a river's flow: gravity.

Quite simply, every river travels from points of higher elevation to points of lower elevation. Most rivers originate in mountains, hills, or other highlands. From there, it's always a long and winding journey to sea level.

Many prominent rivers flow from north to south, which perhaps creates the misconception that all waterways do so. The Mississippi

River and its tributaries flow in a southerly direction as they make their way to the Gulf of Mexico. The Colorado River runs south toward the Gulf of California, and the Rio Grande follows a mostly southerly path.

But there are many major rivers that do not flow north to south. The Amazon flows northeast, and both the Nile and the Rhine head north. The Congo River flaunts convention entirely by flowing almost due north, then cutting a wide corner and going south toward the Atlantic Ocean.

There's a tendency to think of north and south as up and down. This comes from the mapmaking convention of sketching the world with the North Pole at the top of the illustration and the South Pole at the bottom.

But rivers don't follow the conventions of mapmakers. They're downhill racers that will go anywhere gravity takes them.

Q How deep are the oceans?

A First, the boring answer. Going by average depth, the Pacific Ocean is the deepest: 13,740 feet. The Indian and Atlantic oceans are a close second and third at 12,740 and 12,254 feet deep, respectively. The Arctic Ocean is relatively shallow—3,407 feet at its deepest point.

But there's much more to the story. The terrain of the land that lies beneath the oceans is just as varied as the terrain of the higher and drier parts of the globe. The ocean floors have their own mountain ranges—the tallest of which poke through the waves to become islands—and they also have plunging valleys called trenches. These trenches mark the seams at which two of the earth's tectonic plates come together; movement of these plates forces one under the other in a process called subduction. Trenches can be narrow, but they run the entire lengths of the tectonic plates.

Trenches are the deepest parts of the oceans—and the deepest of all the trenches is the Mariana Trench, which is named for its proximity to the Mariana Islands, which are located in the Pacific Ocean between Australia and Japan. The depth of the Mariana Trench varies considerably along its 1,580 miles, but oceanographers have identified one part that's deeper than the others—in other words, the deepest part of the deepest trench in the deepest ocean on the planet. This place is called the Challenger Deep.

The bottom of this valley in the ocean floor is a bone-crushing 35,810 feet deep—almost seven miles below sea level. One way to illustrate this depth is to say that if we wanted to hide Mount Everest—the entire thing—they could toss it into the Challenger Deep. Once it had settled on the bottom, there would still be more than a mile of water between the highest point of the mountain and the ocean's surface.

The Challenger Deep, discovered in 1951, was named for the first vessel to pinpoint the deepest part of the trench: HMS *Challenger II*, manned by Jacques Piccard. The first vessel to plumb these depths was the U.S. Naval submersible *Trieste* in 1960, manned by Piccard and U.S. Navy Lt. Don Walsh. Hydrostatic pressure, caused by the accumulated weight of the water above you, increases as you descend into a body of water. At the bottom of the Challenger Deep, the *Trieste* had to withstand eight tons of pressure per square inch. Bone-crushing, indeed.

Q How do hurricanes get their names?

A Hurricanes are given their names by an agency of the United Nations called the World Meteorological Organization (WMO). The staff of the WMO doesn't spend its days thumbing through baby books trying to pick the perfect name for each newborn storm; instead, it uses a system to assign the names automatically. At the start of each year's stormy season, staffers dust off an alphabetical list of twenty-one names—one for each of the letters except Q, U, X, Y, and Z (these letters are never used due to the scarcity of names that begin with them). As a tropical storm develops, the WMO assigns it the next name on the list, working from A to W and alternating between male and female names. There are six such lists; the WMO rotates through them so that the names repeat every six years.

These lists aren't set in stone, though. If a hurricane causes serious destruction, its name is usually retired. If a hard-hit country requests that a name be replaced, the WMO picks a substitute for that letter of the alphabet. For example, in 2005 Katrina was replaced with Katia, which then appeared in the 2011 rotation. When the WMO picks a new name, the only requirements are that it must be short and distinct, easy to pronounce, generally familiar, and not offensive.

It's not clear who started the tradition of naming hurricanes, but it predates the WMO by hundreds of years. The custom goes back at least to the eighteenth century, when people in the Caribbean began to name storms after the nearest saint's day. For example, the 1876 hurricane that struck Puerto Rico was called Hurricane San Felipe.

A more personalized approach was taken by an Australian meteorologist in the late nineteenth century: He named storms for mythological figures, women, and politicians that he didn't like—but that idea never really caught wind worldwide.

Our current method of naming storms was first formalized by the U.S. National Hurricane Center in 1953, after several naming schemes that used latitude and longitude coordinates and phonetic alphabet signs—the "Able, Baker, Charlie" type of code that's used by the military—proved to be too confusing. At first, storms were given only women's names; this convention was attributed to World War II soldiers who reportedly named storms after their wives. By the 1970s, women understandably felt that it was sexist to link their gender with the destructive forces of tropical storms, so in 1979 the WMO added male names to the lists.

Q Can it rain fish?

A It sure can. While it's a rare occurrence, there are dozens of fishy rainstorms on record. For example, in July 2006, a downpour of pencil-size fish pelted residents of Manna, India. It seems that Mother Nature is one mad scientist.

Long ago, people attributed these fish-storms to the wrath of God or mysterious oceans in the sky. But today, most scientists agree that waterspouts are the actual culprits, though this hypothesis hasn't been definitively proven.

Waterspouts are essentially weak tornadoes that form over bodies of water. Some waterspouts occur in the midst of thunderstorms, similar to the tornadoes that appear on land, but most are fair-weather creations—weaker funnels that can occur even on calm days. Sometimes, the theory goes, these funnels suck up water—

and any creatures that happen to be swimming around in it—from the surface of an ocean or lake. Air currents can keep the fish aloft in the clouds before dropping them onto unsuspecting people who are up to a hundred miles inland.

It's difficult to say how often this freak show occurs, but in some places, it's believed to be an annual event. Residents of Yoro, Honduras, claim that each year between June and July, a big storm leaves the ground covered with tiny fish. These critters may not actually fall from the sky, though; some zoologists theorize that the Yoro fish actually come from underground streams and are stirred up by the rain.

Besides fish, frogs are probably the most common animals to rain from the sky—a plague of falling frogs even makes an appearance in the Old Testament. In July 2005, a mysterious cloud dropped thousands of live frogs onto a small town in Serbia. After fish and frogs, it gets even stranger. Some of the more noteworthy bizarre rainstorms: jellyfish in Bath, England, in 1894; clams in a Philadelphia suburb in 1869; and spiders in Argentina's Salta Province in 2007. We'd be willing to bet that the local weathermen didn't forecast these storms.

Q Why are some bodies of water salty and others are not?

A Actually, all bodies of water contain salt. When the salt concentration is high, as in the oceans, it's obvious to the human tongue. In so-called freshwater lakes and rivers, the concentration is much lower—so low that the tongue can't detect it—but the salt is still there. When oceanographers discuss the salt content (or salinity) of water, they refer to the concentration in terms of parts per thousand. For instance, the saltiest waters in the world—the Red Sea and the Persian Gulf—contain forty pounds of salt for every thousand pounds of water. The major oceans have thirty-five parts per thousand. In contrast, the Great Lakes—the largest freshwater bodies in the world—contain less than an ounce of salt per thousand pounds of water. Your taste buds would have to be mighty sensitive to pick out a salty taste in those waters.

Bodies of water are saltier in regions with higher temperatures; the higher the mercury rises, the more water is evaporated. When water evaporates, salt is left behind. The evaporated water forms clouds that hover high over land and eventually produce rain. The rainwater drains into a river system and picks up salty minerals from the riverbed. When the river rejoins the ocean, it adds more salt to the already salty waters.

This process is what makes Utah's Great Salt Lake so salty. Numerous rivers and streams empty into the lake, carrying with them the same minerals that contribute to an ocean's salinity. Since the Great Salt Lake has no outlet, the minerals that enter have nowhere to go. As a result, some sections of the lake have salinity levels that are eight times higher than those of the saltiest oceans. Salinity in the Great Salt Lake ranges from one hundred and fifty to one hundred and sixty parts per thousand.

You can infer, then, that the saltwater bodies with the lowest salinity would be located in cold regions: the Arctic Ocean, the seas around Antarctica, and the Baltic Sea. The last of these ranges in salinity from five to fifteen parts per thousand.

These bodies of water are constantly being diluted by melting ice and continued precipitation, and they evaporate at much slower rates than hotter oceans and seas; they're always getting more water, but rarely more salt.

Q Does anything live in the Dead Sea?

A The Red Sea isn't really red, and the Black Sea isn't really black—so what are the odds that the Dead Sea is really dead? Then again, if you've ever gone for a dip in the Dead Sea, you'll know that there is at least a shred of truth to the name. The Dead Sea's otherworldly qualities make swimmers buoyant—everyone's doing the "dead man's float."

Located between Israel and Jordan, the Dead Sea is, at thirteen hundred feet below sea level, the lowest surface point on Earth. The very bottom is twenty-three hundred feet below sea level. Water flows into the Dead Sea from the Jordan River, but then it has no place to go, since it's already reached the lowest possible surface point on the planet. The fresh water that flows into the Dead Sea evaporates quickly because of the high temperatures in the desert, and it leaves behind a deposit of minerals.

These minerals have accumulated to make the Dead Sea the pungent stew that it is today. Slightly more than 30 percent of the Dead Sea is comprised of minerals, including sodium chloride, iodine, calcium, potassium, and sulfur. These minerals have been marketed as therapeutic healing products for people with skin conditions; many cosmetic companies have their own line of Dead Sea products.

The Dead Sea is reported to be six times saltier than an ocean, and salt provides buoyancy for swimmers. No form of life could survive in these conditions, right? Not exactly. It's true that every species of fish introduced into this body of water has promptly died, but in 1936, an Israeli scientist found that microscopic pieces of green algae and a few types of bacteria were living in the Dead Sea. So the Dead Sea can't technically be considered dead.

At any rate, it's too late to change the Dead Sea's name—and that's probably just as well. Would the Living Sea Scrolls hold the same intrigue?

Q How come underwater tunnels don't collapse?

A Ever dig a hole at the beach? Just when you think you're finally going to get all the way to China, the walls collapse and your whole hole disappears. (Which, incidentally, is why you should never let children dig at the beach unsupervised.) The length of time your walls will hold up before caving in is known to sandhogs as the "stand-up" time.

Who are sandhogs? They're not little animals burrowing into the sand; they're the men and women who earn their livings in tunnel construction, and they wear the name proudly. Tunnel digging is a high-risk job, and sandhogs have a vested interest in the stand-up time of mud walls, especially when they're digging an underwater tunnel. How exactly can you dig a tunnel through the wet, soggy mud at the bottom of a river without courting certain disaster?

Back in 1818, English engineer Marc Isambard Brunel asked this same question. While strolling on the London docks one day, he noticed a shipworm, a.k.a. *teredo navalis,* boring through some rotting timbers. How did the tiny worm make a tunnel without getting crushed? Close inspection revealed that the worm used its hard, shell-like head as a shield. As the worm ate into the wood like an excavating shovel, its head moved forward, making a tunnel large enough for its body to pass through.

Brunel built himself a cast-iron shield shaped like the head of the worm, only his was two stories high and had several doors for mouths. A worker standing behind the shield would excavate earth through a door. As a hollow space was created in front of the shield, a set of jacks pressed the iron frame forward a few inches at a time, leaving a smooth section of earth where the rim of shield had been. Instantly, bricklayers would get to work, reinforcing this mud wall before it caved in.

As you may guess, this was a slow process. It took Brunel eighteen years to complete a 1,506-foot tunnel beneath London's River Thames. His shield method worked, however, and parts of his tunnel, which opened in 1843, are still in use today.

Though there are several new ways to construct underwater tunnels, including a tunnel–boring machine and prefabricated submersible tubing, the principle behind Brunel's shield continues to be employed in subterranean construction projects.

What is the world's longest underwater tunnel? The Seikan Tunnel, which runs under Japan's Tsugaru Strait, holds the record at 33.49 miles. But in April 2007, Russia announced plans to dig a tunnel to Alaska—a whopping sixty-four miles across the Bering Strait. Impossible? Not if you ask the sandhogs. They'll be there to dig that tunnel through rock, sand, silt, and mud. Just like they've dug all the others.

Q How were the world's great canals built?

A In September 1513, Vasco Núñez de Balboa left the Spanish colony of Darién on the Caribbean coast of the narrow Panamanian isthmus to climb the highest mountain in the area—literally, to see what he could see. Upon reaching the summit, he gazed westward and became the first European to see the eastern shores of the vast Pacific Ocean. From there, he led a party of conquistadores toward his discovery. They labored up and down rugged ridges and hacked their way through relentless, impenetrable jungle, sweating bullets under their metal breastplates the entire way. It took four days for Balboa and his men to complete the 40-mile trek to the beach.

Of course, Balboa's journey would have been much easier if someone had built the Panama Canal beforehand. The construction of great canals such as Suez and Panama cost vast amounts of money and required incredible feats of engineering to link seas and oceans with only a few miles of artificial waterway. But it sure beat sailing thousands of miles around entire continents.

For 2,500 years, civilizations have carved canals through bodies of land to make water transportation easier, faster, and cheaper. History's first great navigational canal was built in Egypt by the Persian emperor Darius I between 510 and 520 B.C., linking the Nile and the Red Sea. A generation later, the Chinese began their reign as the world's greatest canal builders, a distinction they would hold for 1,000 years—until the Europeans began building canals using technology developed centuries earlier during the construction of the world's longest artificial waterway: the 1,100-mile Grand Canal of China.

It's been a long time since a "great" canal was built anywhere in the world. Here's a brief look at the most recent three.

The Erie Canal—Clinton's Big Ditch: At the turn of the 19th century, the United States was bursting at its seams, and Americans were eyeing new areas of settlement west of the Appalachians. But westward overland routes were slow and the cost of moving goods along them was exorbitant. The idea of building a canal linking the Great Lakes with the eastern seaboard as a way of opening the west had been floated since the mid-1700s. It finally became more than wishful thinking in 1817, when construction of the Erie Canal officially began.

Citing its folly and $7-million price tag, detractors labeled the canal "Clinton's Big Ditch" in derisive reference to its biggest proponent, New York governor Dewitt Clinton. When completed in 1825, however, the Erie Canal was hailed as the "Eighth Wonder of the World," cutting 363 miles through thick forest and swamp to link Lake Erie at Buffalo with the Hudson River at Albany. Sadly though, more than 1,000 workers died during its construction, primarily from swamp-borne diseases.

The Erie Canal fulfilled its promise, becoming a favored pathway for the great migration westward, slashing transportation costs a whopping 95 percent, and bringing unprecedented prosperity to the towns along its route.

The Suez Canal—Grand Triumph: The centuries-old dream of a canal linking the Mediterranean and the Red Sea became reality in 1859 when French diplomat Ferdinand de Lesseps stuck the first shovel in the ground to commence building of the Suez Canal.

Over the next ten years, 2.4 million laborers would toil—and 125,000 would die—to move 97 million cubic yards of earth and build a 100-mile Sinai shortcut that made the 10,000-mile sea journey from Europe around Africa to India redundant.

De Lesseps convinced an old friend, Egypt's King Said, to grant him a concession to build and operate the canal for 99 years. French investors eagerly bankrolled three-quarters of the 200 million francs ($50 million) needed for the project. Said had to kick in the rest to keep the project afloat because others, particularly the British, rejected it as financial lunacy—seemingly justified when the canal's final cost rang in at double the original estimate.

The Suez dramatically expanded world trade by significantly reducing sailing time and cost between east and west. De Lesseps had been proven right and was proclaimed the world's greatest canal digger. The British, leery of France's new backdoor to their Indian empire, spent the next 20 years trying to wrest control of the Suez from their imperial rival.

The Panama Canal—Spectacular Failure: When it came time to build the next great canal half a world away in Panama, everyone turned to de Lesseps to dig it.

But here de Lesseps was in over his head. Suez was a walk in the park compared to Panama. In the Suez, flat land at sea level had allowed de Lesseps to build a lockless channel. A canal in Panama, however, would have to slice through the multiple elevations of the Continental Divide.

Beginning in 1880, de Lesseps, ignoring all advice, began a nine-year effort to dig a sea-level canal through the mountains. This futile strategy, combined with financial mismanagement and the death of some 22,000 workers from disease and landslides, killed de Lesseps' scheme. Panama had crushed the hero of Suez.

The Panama Canal—Success! The idea of a Panama canal, however, persevered. In 1903, the United States, under the expansionist, big-stick leadership of Theodore Roosevelt, bought out the French and assumed control of the project. Using raised-lock engineering and disease-control methods that included spraying oil on mosquito breeding grounds to eliminate malaria and yellow fever, the Americans completed the canal in 1914.

The Panama Canal, the last of the world's great canals, made sailing from New York to San Francisco a breeze. A trip that once covered 14,000 miles while circumnavigating South America was now a mere 6,000-mile pleasure cruise.

ANIMALS AROUND THE WORLD

Why don't we ride zebras?

A What's black, white, and red? A zebra enraged at the thought of someone riding on its back.

While the zebra belongs to the horse family, it is kind of like the deranged cousin nobody likes: He gets invited to Thanksgiving out of a sense of familial obligation, but everybody hopes he'll have other plans. It's inevitable that he'll have a few too many drinks, get all emotional, and then start an argument that leaves everyone avoiding eye contact and thinking up excuses for going home early.

Domesticating a zebra is a dicey proposition, though it's been done. For example, it's a real zebra that Hayden Panettiere rides in the 2005 movie *Racing Stripes,* which is about a zebra that aspires to be a racehorse. But the average zebra is far more temperamental than a horse. Zebras spook easily, and they can be exceptionally irritable, especially as they get older. One zebra trainer compared riding the ill-humored beast to "riding a coiled spring."

There have been attempts to create zebra hybrids, such as a "zorse" (a cross between a zebra and a horse) or a "zonkey" (a cross between a zebra and a donkey). But because zebras don't have any particularly useful qualities, such as speed or strength, these hybrids don't have real-world utility beyond the chuckles their whimsical names might elicit.

If you really have a strong urge to "ride a zebra," your best bet might be to head down to the local high school and harass the referees at the football and basketball games. Those zebras may be just as temperamental as the real ones, but they're less likely to kick you in the face.

Q How many animals have yet to be discovered?

A Since about 1.3 million animal species around the planet have been identified and named, you might think that we're down to the last few undiscovered critters by now.

But according to many biologists, we're probably not even 10 percent of the way there. In fact, experts estimate that the planet holds ten million to one hundred million undiscovered plant and animal species, excluding single-celled organisms like bacteria and algae. This estimate is based on the number of species found in examined environments and on the sizes of the areas we have yet to fully investigate.

The broad span of the estimate shows just how little we know about life on Earth. At the heart of the mystery are the oceans and tropical rain forests. More than 70 percent of the planet is underwater. We know that the oceans teem with life, but we've explored only a small fraction of them. The watery realm is like an entire planet unto itself. Biologists haven't examined much of the tropical rain forests, either, but the regions that they have explored have turned up a dizzying variety of life. It's hard to say exactly how many life forms have yet to be discovered, but the majority probably are small invertebrates (animals without backbones).

Insects make up the vast majority of the animal kingdom. There are about nine hundred thousand known varieties, and this number will probably increase significantly as we further explore the rain forests. Terry Erwin is an influential coleopterist—in other words, a beetle guy—who once estimated that the tropics alone could contain thirty million separate insect and arthropod species. This number was based on his examination of forest canopies in South America and Central America, and it suggests that you're on the wrong planet if you hate bugs.

Cataloging all these critters is slow-going. It requires special knowledge to distinguish between similar insect species and to identify different ocean species. It also takes real expertise to know which animals are already on the books and which are not. Qualified experts are in short supply, and they already have a lot on their plates.

In some respects, time is of the essence. Deforestation and climate change are killing off animal and plant species even before they've been discovered. You may not particularly care about wildlife, but these are big losses. The knowledge gained from some of these undiscovered creatures that are on death row could help to cure diseases and, thus, make the world a better place.

Q Why don't penguins and polar bears get frostbite?

A If you spent an afternoon strolling around barefoot at the North Pole or the South Pole, your feet would freeze and—best-case scenario—you'd end the day short a couple of toes.

But polar bears and penguins obviously don't wear boots, and they seem to be fine. What's the deal?

Scientists tell us that the human body evolved to its present state on the toasty African plains, where ice and subzero temperatures are barely imaginable. It's no surprise, then, that your body can't function without significant protection in arctic conditions. When your extremities—your feet and hands—get very cold, your body does something that may seem counterintuitive. As an act of self-preservation, it lets your extremities get colder, constricting the blood vessels that feed those parts in order to conserve heat for the rest of the body. This helps maintain your core temperature in frigid weather, but it wreaks havoc on your hands and feet. The eventual result is frostbite—the tissue dies off. It can also lead to hypothermia, a dangerous drop in overall body temperature, even if the rest of your body is bundled up.

Polar bears don't have this problem because their feet are like natural boots. Their huge paws come equipped with thick pads on the bottom and heavy fur on top. This thick insulation keeps their paws from losing heat rapidly, so there's no need for the bear's body to cut off the blood flow.

The furry, padded-feet approach wouldn't work so well for penguins. For one thing, they need relatively unencumbered feet so that they can swim quickly. Perhaps more importantly, since they're covered in feathers and insulated by layers of fat, their feet are their only means for releasing excess heat when they exert a lot of energy. As a result, penguins evolved highly efficient self-warming feet.

As in humans, the flow of blood to penguins' feet is controlled in order to regulate overall temperature loss. But the blood vessels in their feet are arranged differently than those of humans: The vessels that carry warm blood into the feet are located close to the vessels that take cold blood out of the feet. The warm blood heats up the cold blood that's flowing back into the torso, which prevents their overall body temperature from plummeting. What's more, their bodies are calibrated to keep their feet just a degree or two above freezing; this wards off frostbite.

Penguins have another trick to keep their feet warm. When it gets really cold, a penguin might rest on its heels and tail to keep the majority of its footpads off the ice. As for you? We recommend insulated boots.

Q Are there really alligators in New York City sewers?

A What might be the definitive urban legend goes like this: Back in the early twentieth century, some denizens of Gotham thought that baby alligators made great gifts for kids. Apparently, these knuckleheads couldn't foresee that cute baby alligators would become ugly, adult limb-manglers.

When the gators grew and became dangerous, these New Yorkers flushed the animals into the city's sewer system. There, these warm-climate amphibians found a friendly enough environment, and abandoned alligators, it was said, formed a thriving colony beneath the Manhattan streets.

Fueling the legend was Robert Daley's 1959 book *The World Beneath the City*. According to the book, Teddy May, a former sewer superintendent, claimed that in the 1930s he saw gators as long as two feet in the sewer tunnels. May said that he ordered his charges to kill the reptiles and that it took a few months to complete the job.

Indeed, the 1930s were a golden age for news stories of alligator sightings in and around the city. Oddly, accounts involving the sewer system were rare. Most reports involved surface-level encounters, and the critters in question are believed to have been escaped pets. A February 1935 article in the *New York Times* told of a group of boys who discovered that the manhole into which they were shoveling snow contained an alligator. They used a clothesline to drag the reptile up to the street, then beat it to death with their shovels. Welcome to the Big Apple.

Curiously, no contemporary news coverage of Teddy May's extended alligator hunt can be found. Subsequent published reports painted May as a bit of a raconteur who quite possibly was having some fun at the expense of author Daley. We're not saying you won't encounter some sort of underground wildlife in NYC, but you're more likely to find it on the subway than in the sewers.

Q Did Saint Bernards really rescue people in the Alps?

A Absolutely. In fact, they've rescued about two thousand people since the early eighteenth century. One Saint Bernard named Barry is credited with saving forty lives in the snowy Alps.

The dogs were trained to work in the dangerous Grand-St-Bernard Pass, an ancient route through the Swiss Alps. The eight-thousand-foot-high pass was plagued by deep snow and avalanches, and by thieves who preyed on travelers. Still, the pass was heavily traveled for centuries by farmers and merchants and by the Romans, Huns, and Hannibal's army.

In the tenth century, a priest named Bernard of Menthon came up with the idea of building a hospice on the pass to shelter travelers.

After he died, Bernard became Saint Bernard, patron saint of the Alps.

What about the dogs? The hospice was isolated, so dogs from the valleys were brought up and trained. The breed we call Saint Bernard was common to the region's villages and farms. These dogs were huge and hearty. No one knows when they were first brought to the hospice; the original building and its archives burned in the fifteen hundreds. The hospice was rebuilt, and a 1707 record mentions a dog lost in an avalanche.

From the time of Bernard of Menthon, the monks at the hospice rescued and tended to travelers. Adding dogs was a smart move. Their powerful bodies plowed through snow faster than men could, their noses sniffed out lost travelers, and they had an uncanny ability to find safe passage through the terrain. The dogs made excellent guards, guides, pathfinders, and companions. Contrary to legend, they didn't have little barrels of brandy hanging from their collars. But the dogs would stretch across people they dug out of the snow to warm their bodies, and lick their faces.

Tunnels, roads, and rail lines built in the twentieth century bypass the Saint Bernard hospice, so the pass is not used much anymore—the last documented rescue was in 1897. Few monks now live at the Saint Bernard hospice. In 2004, the hospice made headlines around the world when the monks offered their dogs for sale, with the stipulation that the new owners bring them back to the hospice each year for a visit.

Q Why doesn't a poisonous creature get killed by its own venom?

A Before we get carried away, it's only right to point out the difference between venomous and poisonous creatures. Venomous creatures—such as certain snakes, scorpions, and spiders—produce poison and have developed ways of delivering it (fangs, stingers, etc.). Poisonous creatures—such as dart frogs—also produce poison but have no way of getting it into the bloodstream of the intended victim; the prey has to ingest it.

For the sake of simplicity, we'll focus on the venomous variety. A logical assumption is that all of these creatures have developed immunities to their own toxins, but that's not entirely true. If a snake accidentally bites itself and releases its venom into its body, it will die.

Indeed, the only things that keep the poison in check are the glands that store the poison. You see, these creatures produce venom in special sacs, which are connected to their puncturing tools. These sacs have linings that prevent the venom from escaping to the rest of the creature's body. The sacs aren't unlike your own stomach, which produces all sorts of powerful acids and toxins that would eat you from the inside out if you didn't have a protective stomach lining.

Q Why do we never see baby pigeons?

A If you aren't seeing baby pigeons, you're just looking in the wrong places. Until they're ready to take flight, the young birds hang out in nests while their moms and dads are out socializing.

Pigeons, those head-bobbing city slickers that so often repaint statues and other objects of their "affection," are descended from rock doves, which got their name from their propensity for building their nests on the craggy faces of cliffs. When the rock dove population spills into a metropolitan area—or when the ever-expanding metropolitan area infringes on rock dove territory—the birds build their nests on the unnatural ledges and shelves of tall buildings and bridges.

Baby pigeons, known as squabs, stay in their nests until they are able to fly. Obviously, squabs are as common as adult pigeons. They're just not as visible—unless you're a window washer.

Q Are pandas as huggable as they seem?

A Cute, aren't they? Maybe. But if you want to give a panda a hug, it's best to stick to the plush toy version. Definitely don't mess with Gu Gu—this male panda at the Beijing Zoo has made headlines three times for biting visitors who were foolish enough to climb over the barrier in an attempt to get up close and personal. For all their "aawww" appeal, pandas are still bears. If they feel threatened, they can defend themselves tooth and nail, with their powerful jaws and sharp claws.

Ironically, the big, dark eye patches that endear pandas to people are also their means of signaling aggression toward other animals. Human babies tend to have big eyes relative to the size of their faces, a trait that triggers a protective and nurturing response in most adults. To animals, however, big eyes appear fierce and frightening. When a panda wants to intimidate an enemy, it will hunker down and lower its head to make its eyes seem even larger. If the panda is feeling submissive, it will turn away and hide its eyes with its paws to indicate that it would rather flee than fight. How aggressive are pandas? Male pandas become territorial at around the age of five, when they are ready to mate.

Confrontations between rival suitors aren't always physical; pandas communicate with a wide range of vocalizations, including a harsh bark that young males may employ to rout others from a distance.

An adult male panda typically weighs between two hundred and three hundred pounds, and measures four to five feet from nose to tail. Females are a bit smaller, but can still tilt the scales at close to three hundred pounds.

Not surprisingly, female pandas are most aggressive when they have cubs to protect. They need to be—infant pandas weigh only a few ounces at birth, making them tasty morsels in the wild for wolves, leopards, and birds of prey. Panda cubs remain with their mothers until they reach a roly-poly eighty pounds at approximately eighteen months of age; then they're ready strike out on their own.

Is a panda's bark literally worse than its bite? Fortunately, attacks on humans are rare, but panda-on-panda violence in zoos is a concern. Many attempts to breed pandas in captivity have failed because the couples make war, not love. A study conducted by the Smithsonian Institution revealed that young pandas, like young humans, benefit from social interaction. Bears whose keepers spent time playing with them each day proved most open to potential mates: Those who received little social stimulation tended to be both shyer and more hostile toward would-be wooers.

Though real pandas may not fit the cute 'n' cuddly cartoon stereotype, it doesn't prevent them from keeping their favored status as poster bears for endangered species. After all, what would environmentalists do without those big eyes that implore us to reach into our wallets and save not just the panda, but the rest of the world's wildlife as well?

Q Which fish swims the fastest?

A Fishermen are known for exaggerating stories and boasting about "the one that got away." Not only are those uncaught fish especially big, but they're also usually described as being strong and wily. Excuses abound for why anglers don't manage to reel in these fantastic fish, but you can be sure that it has nothing to do with their skills as fishermen.

And if they were fishing in certain areas of the Indian or Pacific oceans, they might be able to add "too fast" to the list of that wily fish's traits.

As you might imagine, determining the fastest fish on the planet isn't an easy task. Most scientists agree, however, that it is probably the Indo-Pacific sailfish (also known as the Pacific sailfish). It has a huge, dark-blue dorsal fin, a silver-brown body, and a long, pointed bill that it uses to hunt. According to *Guinness World Records,* it can swim sixty-eight miles per hour. The Indo-Pacific sailfish is large in addition to being fast—it usually weighs more than two hundred pounds and can be more than eleven feet long.

Close behind the Indo-Pacific sailfish is the endangered Atlantic bluefin tuna, which has been clocked at up to forty-three miles per hour. The bluefin tuna is even more of a behemoth than the Indo-Pacific sailfish, weighing up to five hundred and fifty pounds.

Good luck trying to reel in either of these fleet fish. Chances are, you'll end up with a new fish tale, about the one that sped away.

Q Which snake toxin can kill you the fastest?

A A big, beautiful, olive-green serpent from the Australian outback is widely regarded as the deadliest (albeit not the most dangerous) snake on the planet. It's a good thing that the inland taipan is such a docile slitherer, because it is said that one of its bites contains enough venom to kill one hundred human adults.

No one knows for sure—when it comes to dangerous snakes, there's no shortage of wild claims (one British television program asserted that a dead rattlesnake could sense heat and would bite a warm object, like a hand)—but even the most sober herpetologists are in awe of the inland taipan. What scientists have verified is that the scaly denizen from Down Under, with its eyes of blackish brown and length that can reach twelve feet, pegs the venom meter.

In point of fact, that meter is a not-particularly-exact ratings method called LD50. It stands for "Lethal Dose, 50 percent" and measures how much venom is required to kill half the members of a test group within twenty-four hours.

In this case, the members of the test group are mice, and inland taipan venom tops the LD50 list with only 0.025 milligram of venom needed per kilogram of mouse weight. The amount of venom in a single dose can kill 250,000 of the little critters. By comparison, the awful-sounding death adder's LD50 of 0.50 milligram per kilogram, which means it takes twenty times as much venom to kill half the test mice, makes this snake relatively benevolent, doing in a mere eleven to twelve thousand mice with the venom of one bite.

Some students of the reptile world say that a human adult who is bitten by an inland taipan would be dead in less than forty-five minutes. Actually, the human would have to be envenomated.

Some killer snakes can sink their fangs into prey without injecting venom; it's what herpetologists call a "dry bite," or what laypeople call a "strong hint."

In any case, once envenomation occurs, the inland taipan's neurotoxin courses through the victim's lymph structure, attacking the nervous system until the muscles that control the lungs are paralyzed. (This is in contrast to the family of venomous snakes that inject a hemotoxin, which attacks the blood system.) Thankfully, not a single human fatality has been attributed to the inland taipan. It's a shy creature that feeds mostly on rats.

This isn't the case with the saw-scaled viper, which is found in India, Sri Lanka, parts of the Middle East, and in Africa north of the equator. A one-and-a-half-foot-long sienna-colored snake with long fangs, it makes a hissing sound when cornered by coiling into an S and rubbing its scales together. Its venom isn't as toxic as that of the inland taipan, but it is dangerous enough. This is the snake you don't want to meet in the grass. The saw-scaled viper, which is said to be responsible for more human deaths per year than any other snake, tops the list of dangerous serpents.

Q Which is the world's most poisonous creature?

A There is no way to answer this question scientifically. To do so, in the words of Otter from *Animal House,* "would take years and cost thousands of lives." Mouse lives, at least; you would have to be willing to sacrifice a few million rodents to test the toxins from the known deadly animals and then extrapolate from those findings to humans.

But we can speculate on such a joyously gruesome topic. Many have suggested that the winner would be *Chironex fleckeri,* one of several types of box jellyfish, which technically aren't even jellyfish. (They are cubozoans; jellyfish are scyphozoans.) This Australian marine stinger is particularly deadly; its venom can kill sixty humans in three minutes.

Not only are these bad boys lethal, having accounted for up to one hundred human deaths in the past century—they're also freaky. As Dan Nilsson of the University of Lund in Sweden puts it, "These are fantastic creatures with twenty-four eyes, four parallel brains, and sixty arseholes."

Let's talk about other potential candidates, if only to satisfy our curiosity. *Discovery Magazine* suggests that the poison arrow frog and certain salamanders as contenders for the most-poisonous title. "Just two micrograms of toxin from the poison arrow frog is enough to kill a human," the magazine states, then notes for emphasis that the ink in the period of a sentence is three times that volume.

Discovery Magazine also says that the ugly stonefish—probably the world's most venomous fish—and the inland taipan snake mentioned in the last essay should be given consideration. But tree-huggers could note, with some objective evidence, that the most poisonous animal is man. Dioxins, among many other deadly poisons, didn't exist until we invented them.

Do animals still get stuck in the La Brea Tar Pits?

Yes, animals still get stuck in the La Brea Tar Pits. They're much smaller than they used to be, though. None of your saber-toothed cats or woolly mammoths—we're talking lizards and birds, mostly.

Tar pits are created when crude oil seeps through a fissure in the earth's crust. When oil hits the surface, the less dense elements of the oil evaporate, leaving behind a gooey, sticky substance known as asphalt, or tar.

La Brea is a Spanish name that translates literally to "the tar." These famous pits are located in Hancock Park in Los Angeles; they constitute the only active archeological excavation site to be situated in a major metropolitan area.

The pits are tended by staff members of the George C. Page Museum, which is nearby. It is at the museum that the fossils currently being excavated from the pits are cleaned and examined. Not only do the staff members get to parse through the well-preserved remnants of prehistory, but they also sometimes witness the natural process by which these remnants are preserved. An average of ten animals every thirty years get trapped in the pits.

The tar pits work like a large-scale glue trap. If an animal lets just one paw hit the surface of the asphalt, it sticks (especially on warm days, when the asphalt is at its stickiest). In its frenzy to free itself, the animal gets more stuck. Eventually, its nose and mouth will be covered, and then it's all over—just one more carcass for scientists to excavate.

The La Brea Tar Pits are forty thousand years old. Since the early twentieth century, scientists have uncovered the remains of more than 600 different species of plants and animals there. Because of the preservative qualities of the tar, and because most of the skeletons are complete, the discoveries from the tar pits have made La Brea an indispensable resource for the scientific community.

As for the animals that get stuck in the pits? They're not as lucky.

Q How come hibernating animals don't starve?

A They binge, then go comatose. Does this sound like something that's happened to you after you polished off a quart of Häagen-Dazs? Can you imagine that food coma lasting all winter?

Animals that hibernate have triggers that warn them to glut themselves for the winter ahead. As the days get shorter and colder, the critters' internal clocks—which mark time through fluctuations of hormones, neurotransmitters, and amino acids—tell them to fill up and shut down. Bingeing is important; if these creatures don't build up enough fat, they won't survive. The fat that they store for hibernation is brown (rather than white, like human body fat) and collects near the brain, heart, and lungs.

Animals have a number of reasons for hibernating. Cold-blooded creatures such as snakes and turtles adjust their body temperatures according to the weather; in winter, their blood runs so cold that many of their bodily functions essentially stop. Warm-blooded rodents can more easily survive the extreme chills of winter, but they have a different problem: finding food. They most likely developed their ability to hibernate as a way of surviving winter's dearth of munchies.

After an animal has heeded the biological call to pig out, its metabolism starts to slow down. As it hibernates, some bodily functions—digestion, the immune system—shut down altogether. Its heartbeat slows to ten or fewer beats per minute, and its senses stop registering sounds and smells. The animal's body consumes much less fuel than normal—its metabolism can be as low as 1 percent of its normal rate. The stored fat, then, is enough to satisfy the minimal demands of the animal's body, provided the creature found enough to eat in the fall and is otherwise healthy.

It can take hours or even days for the animal's body temperature to rise back to normal after it awakens from hibernation. But time is of the essence—the beast desperately needs water, and thirst drives it out of its nest. However, the animal is groggy and slow of foot—it walks like a drunk—so it can be easy prey if it does not get hydrated quickly.

Which animals hibernate? Small ones, mostly—cold-blooded and warm-blooded critters alike. The first category includes snakes, lizards, frogs, and tortoises; the second includes dormice, hedgehogs, skunks, and bats.

But what about the bear, the animal that is most closely associated with hibernation? Whether a bear hibernates has actually been a source of scientific controversy over the years. In the United States circa 1950, the story of the hibernating bear was told with confidence and abandon. Schoolchildren from coast to coast knew of the sleepy bear that, come cold temperatures and snow, escaped into a cave for months of deep respite. With the arrival of spring and warmer weather, the bear would emerge from the cave to search for food and frolic in a nearby stream.

And then came the scientists with their sophisticated shiny metal objects, which they used to measure metabolism, temperature, and oxidation. In the 1960s, '70s, and '80s, many such scientists concluded that bears do not hibernate. The logic went something like this: When animals hibernate, their body temperature drops. Smaller mammals that hibernate can drop their body temperature below freezing.

Bears, however, drop their body temperature by only 10 to 15 degrees. Further, whereas some smaller mammals cannot be easily awakened during hibernation, a hibernating bear can be stirred from its sleep with relatively little effort. The conclusion was that bears do not hibernate.

This created quite a stir in the scientific community. If bears aren't hibernating, what exactly are they doing? A replacement theory was the concept of "torpor," a biological state in which animals lower their metabolic rates, but generally for shorter periods of time. Torpor is considered to be less of a "deep sleep" than hibernation and is seen in birds, rodents, insectivores, and marsupials.

Yet the torpor argument came with its own set of problems. In many respects, the hibernation period of a bear is actually deeper than that of other hibernating species. Although rodents and other small mammals drastically reduce their body temperatures during hibernation, they wake every few days in order to eat and urinate. Some species of bears are able to go six to eight months without eating, urinating, defecating, or fully waking.

Further, it is because of their large size that bears do not drastically reduce their body temperatures during hibernation. In the face of this confusion, words such as denning and dormancy were coined to describe the habits of bears.

Whether the physical inactivity of an animal is labeled hibernation, torpor, or denning, the purpose of these prolonged states is to conserve energy in the face of food scarcity or uncomfortable temperatures.

The particular strategy a species takes in these endeavors varies greatly and depends on its environment. In the colder northern regions of North America, where food is unavailable for the long stretch of winter, black bears hibernate for several months. In the Arctic Circle, where food supplies are unpredictable, polar bears can go into hibernation at any time of year. Only female polar bears hibernate, no doubt because of the elevated energy requirements of pregnancy, birth, and feeding the young.

Scientists moved back into the "pro-hibernation" camp after the discovery of a lemur that hibernates in the tropics, apparently to save energy in the face of heat and food scarcity (the word estivation refers to species that hibernate in warm temperatures). It was traditionally thought that hibernation served as an "escape" from the cold, but the energy-saving behavior of the lemur's hibernation demonstrated that the process is about lowering metabolism in the face of environmental stressors that vary from species to species. "Hibernate" is now accepted as a broader phrase that refers to a reduction in metabolism for prolonged periods of time—meaning it is once again safe to tell the tale of the hibernating bear. Still, if you're ever taking a peaceful nature walk on a sunny winter morn, beware. A bear might be out there.

Q Which animal has the longest lifespan?

A This is trickier to answer than you might think. After all, you can't just go down to the animal retirement home and see which one wears its pants the highest.

For starters, you have to define what a lifespan is, which isn't necessarily a straightforward proposition. Consider the humble amoeba. These protozoa are single-celled animals that reproduce asexually with a process called binary fission. In other words, they don't give birth—they just split in two. This raises the question: Does the lifespan of an individual protozoan continue when it splits apart? If so, then any particular member of a protozoan species has lived for as long as the species itself—which would be millions of years.

Coral presents a similarly head-scratching question. A coral formation is made up of lots of tiny polyps, each with its own "mouth" that can capture plankton and is anchored to a sort of skeleton that is left behind by dead polyps. If you think of a chunk, or "head," of coral as an animal, then coral's lifespan is six thousand years or longer. But if you think of coral as a colony of many tiny, discrete animals—the polyps—then it's not remotely in the running for the title of oldest creature.

Moving on to more animal-like animals, one possibility is *Turritopsis nutricula,* a peculiar type of hydrozoan. Like many other hydrozoans, this species goes through two life stages—it begins as a polyp before eventually turning into a jellyfish. But what makes this particular hydrozoan unique is its ability to turn back into a polyp and start the life cycle all over again. In theory, an individual polyp/jellyfish could switch back and forth indefinitely, attaining a kind of immortality. There's no proof of age for any individual *Turritopsis nutricula,* however.

But enough with the semantics and theories. The oldest single animal on record is a quahog clam that was found off the coast of Iceland in 2006. The researchers who discovered it couldn't check its license, of course, but they estimated its age by counting its growth rings, which form roughly once a year, as more than 400 years old. In 2013, a new estimate showed the clam (now deceased) as more than 500 years old. Unless these scientists happened to come across the Methuselah of the clam world, it's likely there are some that are even older. Ol' Grandpa Quahog's age was only an estimate, as is any age assigned to an animal in the wild. To be sure, you have to keep tabs on an animal, which means that it needs to be in captivity.

And the animal that lived the longest in captivity was probably an Aldabra tortoise named Addwaita. According to officials at the Calcutta zoo in which it was kept, Addwaita was originally given as a gift to Lord Robert Clive of the East India Company in the 1760s. It died in 2006; according to the zoo, it was about 250 years old. So, if you're thinking of getting a giant tortoise, clam, or *Turritopsis nutricula* as a pet, make sure that you have lots of kids. Future generations will be taking care of your animal companion long after you are gone.

Q How fast is a snail's pace?

A The word slow hardly begins to cover it. These animals make all others look like Speedy Gonzales. Next to the snail, tortoises look like hares, and hares look like bolts of furry brown lightning.

Which brings a bad joke to mind: What did the snail riding on the tortoise's back say?

Whee!

Garden snails have a top speed of about 0.03 mile per hour, according to *The World Almanac and Book of Facts.* However, snails observed in a championship race in London took the thirteen-inch course at a much slower rate—presumably because snails lack ambition when it comes to competition. To really get a snail moving, one would have to make the snail think its life was in jeopardy. Maybe the racing snails' owners should be hovering behind the starting line wearing feathered wings and pointed beaks, cawing instead of cheering.

The current record holder of the London race, the Guinness Gastropod Championship, is a snail named Archie, who made the trek in 1995 in two minutes and twenty seconds. This calculates to 0.0053 mile per hour. At that rate, a snail might cover a yard in 6.4 minutes. If he kept going, he might make a mile in a little less than eight days. In the time it takes you to watch a movie, your pet snail might travel about fifty-six feet. You could watch a complete trilogy, and your snail might not even make it out of the house. Put your pet snail on the ground and forget about him—he'll be right around where you left him when you get back.

So long as no one steps on him, that is.

How much wood can a woodchuck chuck?

"How much wood could a woodchuck chuck, if a woodchuck could chuck wood?"

This classic tongue twister has been part of the English lexicon for ages. But has anybody really thought about what it means? Has anybody even seen a woodchuck chucking wood? Or chucking anything, for that matter?

Part of the confusion lies in the origin of the word woodchuck. A woodchuck *(Marmota monax)* is, in fact, the same thing as a groundhog. In the Appalachians, it's known as a whistle pig. According to etymologists, the word woodchuck is probably derived from early colonial British settlers who bastardized wuchak, the local Native American word for groundhog. Because many early Americans couldn't be bothered to learn languages other than English (sort of like present-day Americans), they simply transformed the Algonquian word into one that sounded like an English word.

That the name made absolutely no sense mattered little to these settlers, who were far more concerned with issues like starvation and massive epidemics of fatal illnesses.

Still, the question remains. What if woodchucks could chuck wood? Not surprisingly, there is little research on the topic. Indeed, no studies as of yet have proved that woodchucks are even capable of chucking wood, though there is ample evidence that woodchucks enjoy gnawing through wood when they encounter it.

There is, however, one thing that woodchucks are adept at chucking: dirt. The average woodchuck is quite a burrower, building complicated underground bunkers. These tunnels have been known to reach more than forty-five feet in length with a depth of several feet. Based on these measurements, one woodchuck expert determined that if the displaced dirt in a typical burrow was replaced with wood, the average whistle pig might be able chuck about seven hundred pounds of it.

In the end, the best answer is probably provided by the rhyme itself. "How much wood could a woodchuck chuck, if a woodchuck could chuck wood? A woodchuck would chuck all the wood he could, if a woodchuck could chuck wood."

Which would probably be none.

Q What eats sharks?

A One of the most feared animals in the world, the shark has a reputation for being a people killer, ruthlessly nibbling on a leg or an arm just to see how it tastes. But in the shark vs. people debate, guess who loses? Yup, sharks. We eat way more of them than they do of us. And we aren't the only ones partaking in their sharkliciousness.

For the most part, the big predator sharks are in a pretty cushy position ecologically. As apex predators, they get to do the eating without all that pesky struggle to keep from being eaten. They are important to the ecosystem because they keep everything below them in check so there are no detrimental population booms. For example, sharks eat sea lions, which eat mollusks. If no one ate sea lions, they'd thrive and eat all the mollusks.

So if sharks are apex predators (so are humans, by the way), they aren't ever eaten, right? Wrong. Sometimes a shark gets a hankering for an extra-special treat: another shark.

Tiger sharks start eating other sharks in the womb: Embryonic tiger sharks will eat their less-developed brothers and sisters. This tradition of eating fellow tiger sharks continues through adulthood. And great white sharks have been found with four- to seven-foot-long sharks in their stomachs, eaten whole.

There's also what is called a feeding frenzy. What generally happens is that an unusual prey (shipwreck survivors, for example) presents itself and attracts local sharks, which devour the unexpected meal. The sharks get so worked up from the frenzied feeding, they might turn on each other.

Orcas and crocodiles have also been known to snack on shark when the opportunity presents itself. Note that both orca and crocodiles are also apex predators. So while there are no seafaring animals that live on shark and shark alone, sharks aren't totally safe.

Finally, there's that irksome group of animals known as humans. Many people who reside in Asia regularly partake of shark fin soup, among other dishes prepared with shark ingredients.

Through over-fishing, humans reduced the shortfin mako's population in the Atlantic Ocean by 68 percent between 1978 and 1994.

Even with all this crazy shark-eating, it's a good bet a sea lion or mackerel would happily trade places with the apex predator any day of the week.

GEOGRAPHY GRAB BAG

Q Which ancient Greek first drew a map of the world?

A The Greek philosopher Anaximander is one of the greatest thinkers of all time and, with his wide range of interests and brilliant mind, fits the mold of a "Renaissance Man." Of course, he was born 2,000 years before the Renaissance.

Anaximander was the first philosopher in history to have written down his work—perhaps that is also why he's known as such a groundbreaker. Unfortunately, even though he produced and recorded the work, for the most part it has not survived. We mostly know of Anaximander through doxographers, or writers who document the beliefs, thoughts, and theories of their predecessors. In Anaximander's case, Aristotle and Plato have told us most of what we know about his work.

Born in ancient Greece in the seventh century B.C., Anaximander came from Miletus, a city in Ionia, which is now the western coast of Turkey. This area was a cultural enclave known for its progressive views on philosophy and art—it paved the way for the brilliant artistic development of Athens in the fifth century B.C.

A true rationalist, Anaximander boldly questioned the myths, the heavens, and the existence of the gods themselves. He wanted to devise natural explanations for phenomena that had previously been assumed to be supernatural.

As founder of the science of astronomy, he was credited with building the first gnomon, or perpendicular sundial, which he based on the early work of the Babylonians and their divisions of the days.

Breaking new ground in geography as well, Anaximander has also been credited as being the first cartographer to draw the entire inhabited world known to the Greeks. The map was likely circular, and a river called Ocean surrounded the land. The Mediterranean Sea appeared in the middle of the map, and the land was divided into two halves, one called Europe and the other, Southern Asia. What was assumed to be the habitable world consisted of two small strips of land to the north and south of the Mediterranean Sea.

The accomplishment of this map is far more significant than it might originally appear. Firstly, it could be used to improve navigation and trade. But secondly, and perhaps more importantly, Anaximander thought that by displaying the lay of the land, so to speak, and demonstrating which nations and people were where, he might be able to convince the Ionic city-states to form a federation to push away outside threats.

Watching the horizon, Anaximander concluded that Earth was cylindrical, its diameter being three times its height, with man living on the top. He also thought that Earth floated free in the center of the universe, unsupported by pillars or water, as had commonly been believed at the time. "Earth didn't fall," Aristotle recounted, "because it was at equal distances from the extremes and needed not move in any particular direction since it is impossible to move in opposite directions at the same time."

These were amazing and progressive ideas primarily for one simple reason: They were not based on things that Anaximander could have observed but, instead, were the result of conclusions he reached through rational thought. This is the first known example of an argument based on the principle of sufficient reason, rather than one of myth.

By boldly speculating about the universe, Anaximander molded the direction of science, physics, and philosophy. The idea that nature is ruled by laws just like those in human societies, and anything that disturbs the balance of nature does not last long, should make man pause and think...just as Anaximander did.

Q Can you buy your own island?

A Private islands used to be exclusively for celebrities and ex-dictators. Now, it seems that anyone can own one. A quick Google search will turn up thousands of ads from real estate agents who are eager to sell you the island of your dreams. "Hot Private Islands!" touts one. "Don't just buy a home," another declares. "Buy an island!"

How deep do your pockets have to be? According to the *Wall Street Journal,* you'll pay anywhere from $150,000 for a small island off the North America coast to forty million dollars for a Caribbean retreat. Surprisingly, islands may be cheaper than shorefront property. In Florida, a lux beach house goes for about $3.5 million, while a coastal island sets you back a mere $2.6 million.

However, if you're thinking about emptying your bank account and parachuting into paradise, look before you leap. That's the gist of Cheyenne Morrison's advice. Morrison, a self-described "islomaniac" and a former real estate agent specializing in private islands, keeps a blog called (what else?) "The Private Islands Blog." What does this PI guru think you need to know if you're a serious island buyer?

First, what you see may not be what you get. High tides submerge many small islands, so make sure you visit your potential purchase at different times, just to make sure you won't be treading water for twelve hours a day. Hurricane season, which stretches from June to the end of November, can also put a dent in your wallet and your vacation plans. And even in good weather, the cost of transporting goods to an island can make living there significantly more expensive than living on the mainland.

While it sounds nice to be ruler of your own realm, every one of the world's approximately six hundred thousand islands falls under the jurisdiction of a national government. Whether, what, and how much you can build may be subject to a myriad of complex regulations, as may your source of electric power, your water supply, and your use of any natural resources.

Nor are all islands that are for sale unoccupied. You could wind up as the local authority over a town or village. If you're as wealthy as Malcolm Forbes, no problem. During the 1970s, the publishing magnate subsidized the education of the entire population of his Fijian Island, Laucala. Playing mayor, though, may not appeal to the average retiree. Uninvited guests, a.k.a. squatters, can bring headaches, too. In Costa Rica, anyone who lives on an island for more than three months can petition for permanent residence. Private property, as it turns out, is not so private in many countries, which is another issue to consider before signing the deed.

So is it worth it? Every year, approximately four hundred islands hit the marketplace. Some buyers find the homes of their dreams, while others only have a good story when they're on the selling end of the process. Can't decide? You can always go back to browsing cyber-listings. It doesn't cost anything to click. Sometimes that's the best way to dream.

Q Do a landowner's rights extend all the way to the earth's core?

A Like most legal issues, it depends on the paperwork. Deeds to land are all written differently.

The simplest form of ownership is appropriately named "fee simple" and entitles you to the land and everything above and below it. Yes, that technically means you would own all the land below you (mineral rights) and all the air above you (surface rights). But it's not that cut-and-dried.

For instance, back before planes became a reality, there was no need to restrict your ownership of airspace. People who owned land also owned all the air above it and could use it as they pleased. Once air travel came about, the powers-that-be realized some guidelines needed to be established, so nowadays, ownership is generally restricted to the airspace that landowners can reasonably use or occupy.

In the United States, the Federal Aviation Administration is in charge of all the airspace "above the minimum altitudes of flight," which means you can't really make a stink about aircraft flying far overhead.

Today, the same rules usually apply to underground ownership: Your ownership only goes as deep as you reasonably need. You may or may not have "mineral rights," which would entitle you to any mineral deposits underneath your property. If someone else owns these rights, he or she can come in and dig up the minerals.

So if you want to attempt to dig to the center of the earth, there may not be anyone who can stop you. Or there may be. The only way to know for sure is to pull out your glasses and take a hard look the fine print in your deed.

Q Are there more people alive than dead?

A It would be comforting if this were true, since it would give us an advantage in the event of a zombie uprising.

But no, this oft-cited statistic is wrong; in fact, the dead outnumber the living by a huge margin. The popular idea that there are more people alive than dead took hold in the 1970s and was widely disseminated. It's not clear, however, who first made this faulty claim—but it is faulty.

In 1995, a demographer named Carl Haub got out his calculator and started crunching the numbers to tally the living/dead split. You'll get a different estimate for the age of the human race depending on whom you ask, but Haub went with the United Nations' official estimate that we go back about fifty thousand years. He started his count with two people, a man and a woman, living the high life back in 48,000 BC. And he kept counting...and counting...and counting.

Based on historical population totals and growth rates, he calculated that, as of 2002, 106 billion people had been born in human history. Earth's living population at the time was 6.2 billion, or about 6 percent of that total. Other estimates put the number of dead at only sixty billion, but that still constitutes a sizable lead over the number of people now living.

Will the breathers ever overtake the living-impaired? Don't bet on it. The population growth rate has been dropping over the past forty years, and the United Nations estimates that sometime after 2200, the human race will stabilize at a population of approximately ten billion.

To overtake the dead, our numbers would have to grow to about one hundred billion, but we would likely deplete all of our planet's natural resources before our population could grow that large. Nevertheless, if we solve the resources problem (by colonizing other planets, say) and the tricky little death problem (by turning into cyborgs, say), we could make it happen.

If there's an apocalyptic living vs. dead battle anytime soon, the hopelessly outnumbered living will have to depend on superior technology to win the day. That wouldn't be a problem, though. Remember, most of those dead people never even heard of electricity and firearms when they were alive.

Q What would happen if everyone flushed the toilet at the same time?

A Don't let this keep you up at night. If the President of the United States declared a mandatory national potty break, we wouldn't see our pipes bursting or sewage flowing in the street.

Let's review Sewage 101. When you flush your toilet, the water and waste flow through a small pipe that leads to a wider pipe that runs out of your house. If you have a septic tank, your waste's fantastic voyage ends there—in a big concrete tub buried under your yard. But if your pipes are connected to a city sewer system, the waste still has a ways to go: The pipe from your house leads to a bigger pipe that drains the commodes of your entire neighborhood;

that pipe, in turn, leads to a bigger pipe that connects a bunch of neighborhoods, which leads to a bigger pipe, and so on, in a network that contains miles and miles of pipe.

Eventually, all the waste reaches the sewage treatment plant. The pipes slant steadily downward toward the treatment plant so that gravity keeps everything moving. Where the terrain makes this impossible, cities set up pumps that move the sewage uphill. And fortunately for us, the pipes at each stage are large enough to accommodate the unpleasant ooze that results from all of the flushing, bathing, and dishwashing that goes on in the connected households, even at peak usage times.

It's true that if an entire city got together and really tried, it could overwhelm its sewage system—pumping stations and treatment plants can only deal with so much water at a time, and pipes have a fixed capacity, too. Sewage would overflow from manholes and eventually come up through everyone's drains. But toilet flushes alone aren't enough to wreak such horrific havoc.

A flush typically uses between 1.5 and 3.5 gallons of water. (Federal law mandates that no new toilet can use more than 1.6 gallons of water per flush, but older toilets use more.) There's plenty of room for that amount of water in the pipes that lead out of your house—even if you flush all your toilets at once. Similarly, even if an entire city were to flush as one, there would still be space to spare.

To create a true river of slime, you and your neighbors would have to run your showers, dishwashers, and washing machines continuously; you could even add a flush or two for good measure. (Note that every area's sewer system is self-contained; flushing in unison all over the world wouldn't make things worse in any particular city.)

Still, the fear that such a calamity could occur has inspired some persistent urban legends, like the so-called Super Bowl Flush. In 1984, a water main in Salt Lake City broke during halftime of the big game, and reporters initially said that it was the result of a mass rush to the can. In reality, it was just a coincidence—mains had been breaking regularly in Salt Lake City at the time. But the story stuck, so when the Super Bowl approaches, you're bound to hear that it's best to stagger your flushes at halftime for the greater good.

Q How come Esperanto never caught on?

A Esperanto has fallen short of the hopes of its creator, L.L. Zamenhof, but it has by no means been a flop. When the Russian-born Zamenhof unveiled Esperanto in 1887, he envisioned it as a flexible world language that would become the shared tongue of governments everywhere and would promote peace and understanding. Although Esperanto never became that pervasive, for decades it has been the most used "model" or "constructed" language in the world, with estimates of current users ranging as high as two million. That's more than the speakers of many natural languages, such as those spoken by Native Americans, aboriginal populations on other continents, and European minority peoples.

According to estimates, there are a few hundred to about a thousand native speakers of Esperanto—folks whose parents taught them the language as a baby. Among these is gazillionaire Hungarian financier, activist, and philanthropist George Soros. Although Soros can stay in—or buy, for crying out loud—any hotel in the world, he can also use his Esperanto to secure lodgings with any other Esperantist in roughly ninety countries, one of the language's endearing features.

Tens of thousands of books have been published in Esperanto, including original and translated works; there are Esperanto television and radio broadcasts, magazines, and an annual world congress that attracts an average of two thousand attendees. Its proponents say it is up to twenty times easier to learn than other languages.

So why didn't Esperanto succeed on Zamenhof's terms? There are several reasons:

• Given the disparity of languages in the world, it is impossible to construct a vocabulary and grammar that doesn't pose serious challenges to someone.

• The language's sounds are too similar to Zamenof's native Belarussian, making pronunciation hard for many. Culturally speaking, it is European in its vocabulary and semantics.

• The vocabulary is unnecessarily large, due to the constant additions of new word roots rather than new words being based on old ones.

• Esperanto and its speakers have been the subject of persecution. It was outlawed in communist Russia until 1956, and Esperanto speakers have been killed under totalitarian regimes. Hitler claimed it could become the language of an international Jewish conspiracy, which was a testament to both Esperanto's success and Hitler's insanity.

But the biggest strike against the language is that it sprung from no shared natural culture. To grow large, a language needs a considerable group of people speaking together daily and developing close associations among their shared experiences and shared words. Languages are an outgrowth of human behavior and history, not the result of well-intentioned intellectual efforts.

Put it this way: Any language whose biggest gabfest is an annual world congress isn't going to take over the world.

Still, Esperanto was and is a success. Consider this: Around the time Zamenhof was inventing Esperanto, a German priest, who was acting on something God told him in a dream, invented Volapük. The language got off to a fast start, but today it's estimated that only a few dozen people speak Volapük. Now that's a lonely world congress.

Q If lightning terrifies you, where in the world should you avoid?

A Lightning frightens plenty of people, as Shakespeare pointed out in *King Lear*: "To stand against the deep dreadbolted thunder?/In the most terrible and nimble stroke/Of quick, cross lightning?"

Statistics provided by Britain's Royal Aeronautical Society back up the bard's fear. There are about twenty-four thousand lightning fatalities worldwide each year.

What can you do to avoid becoming one of these statistics? Well, don't play golf or soccer, for a start. Participants in those two sports are in particular danger because they're out in open fields. Multiple-death incidents have been reported on soccer pitches in Malawi, Indonesia, Malaysia, Colombia, Honduras, and Guatemala since 1993. Other lightning-avoidance strategies include not hanging around outdoors in threatening weather and staying off your landline telephone. Landline phone use is the leading cause of indoor lightning injuries in the United States—but it's safe to use a cordless phone or a cell phone.

And for your next family vacation, find a locale that is light on lightning. Since lightning is produced by the collision of hot and cold air, the ideal spot has a basically steady climate. Find a spot without wild weather extremes, either someplace cool year-round (like eastern Russia, northern and western Canada, or Alaska) or consistently warm and mild (Hawaii, if you can afford it). Alaska and Hawaii are the only states that haven't had a lightning fatality since 1959.

Then there are the spots to avoid, starting with central Africa. According to *Extreme Weather: A Guide and Record Book,* there are eight places in the world that endure more than two hundred thunderstorm days per year, and six are in central Africa. Kamembe, Rwanda, in east-central Africa, is third on this dubious list, with an average of 221 thunderstorm days. According to the National Lightning Safety Institute, Kamembe endures an incredible 82.7 lightning strikes per square kilometer each year, an average of sixteen more strikes than any other area in the world.

In the United States, don't rush to these general areas, according to the authors of Extreme Weather: the Gulf Coast, the Florida peninsula, and the peak of the front range of the Rocky Mountains. Florida easily tops all states in lightning deaths, with an average of ten per year.

Don't linger out in the countryside; urban centers are safer. Authorities attribute the significant decrease in lightning-related deaths and injuries in the United States since 1900 to the country's mass migration from rural areas to the cities.

As for Shakespeare and lightning, he was pretty darned safe, despite what he wrote. Britain's Met Office notes that the country averages three lightning deaths per year. That's one per twenty million residents—among the lowest rates in the world.

Q Are there cultures in which a woman can have multiple husbands?

A When someone mentions polygamy, we usually think of certain Mormon fundamentalists or Muslim sects that allow a man to have multiple wives. Anthropologists use the term "polygamy" to refer to any marriage system that involves more than two people. The "one husband, many wives" form is called "polygyny." The "one wife, multiple husbands" version is called "polyandry." And if anthropologists have a special word for it, it must exist somewhere.

Polyandry is, however, exceedingly rare. Only a few cultures continue to practice it today, and polyandry is gradually being eroded by more modern ideas of love and marriage. The strongholds of polyandry are Tibet, Nepal, and certain parts of India, and it is also practiced in Sri Lanka. Other cultures were polyandrous in the past, though it was never widespread.

In many cases, this marriage practice takes the form of fraternal polyandry, in which one woman is married to several brothers. There may be a primary husband, the eldest brother or the first one she married. If additional brothers are born after the marriage, they usually become the woman's husbands as well.

The reasons for polyandry are typically more economic than religious. In the areas where it is practiced, life is difficult and poverty is rampant. If a family divided its property among all the brothers, no one would have enough land to survive on through farming and herding. Keeping all the brothers as part of one family keeps the familial plot in one piece. The herding lifestyle also means one or more brothers are often away tending the livestock for extended periods, so the other husbands can stay at home, protect the family, and tend the farm. Where resources are so limited, polyandry also serves as a form of birth control, since the wife can only get pregnant once every nine months no matter how many husbands she has. No one is ever sure which father sired which child, so each tends to treat all of the kids as if they are his own.

Q How do generations get their names?

A In The Who's 1965 song "My Generation," lead singer Roger Daltrey famously declared that he hoped he'd die before he got old. But what exactly was Daltrey's generation? For starters, it didn't die. Instead, it got old, giving us, among other mediocrities, minivans along the way. We know the people from this generation as Baby Boomers.

Agreeing on a name for a generation is tricky business. Equally tricky is figuring out exactly who belongs to which generation. (There is no official Council of Generation Naming.) If it's a problem, though, it's a fairly recent one. It is only with the rise of en masse self-consciousness—and rampant narcissism—that people have even cared about naming generations.

Historically, generations have gotten their names from literary sources. One of the first generations to name itself was the Generation of '98, a movement of Spanish artists and writers who pointed to the Spanish-American War of 1898 as a break from an artistic and political past. The generational term Baby Boomer was coined in a 1974 *Time* magazine article about Bob Dylan. And Generation X, though around for some time beforehand, was cemented in the popular mind by Douglas Coupland's eponymous 1991 novel.

Possibly the most famous and influential—and certainly the most romantic—generation of the twentieth century was known as the Lost Generation. The term is attributed to writer Gertrude Stein, who used it to describe disillusioned World War I veterans. Stein's cohorts in Paris, among them Ernest Hemingway, embraced the term. Of course, critics are quick to point out that a few artists and expatriates traipsing about the Left Bank hardly qualify as an entire generation. Which points to the futility of such an exercise in generalization in the first place.

What do maps and jigsaw puzzles have to do with each other?

Puzzled over the origin of the jigsaw? We've put the pieces together!

In the 1760s, engraver and cartographer John Spilsbury cut a wooden map of the British Empire into little pieces. Reassembling the map from the parts, he believed, would teach aristocratic schoolchildren the geographic location of imperial possessions and prepare them for their eventual role as governors. Spilsbury called his invention "Dissected Maps."

Spilsbury's "Dissected Maps" were soon popular among the wealthy classes. By the end of the 19th century, however, these "puzzles" functioned largely as an amusement. The early decades of the 20th century were the heyday of the puzzle's popularity in the United States. The first American business to produce jigsaw puzzles was Parker Brothers; the company launched its hand-cut wooden "Pastime Puzzles" in 1908. Milton Bradley followed suit with its "Premier Jig Saw Puzzles," so named because the picture, attached to a thin wooden board, was cut into curved and irregular pieces with a jigsaw. As the Great Depression came to an end in the late 1930s, inexpensive cardboard puzzles were produced at prices nearly anyone could afford.

For more than a century, jigsaw puzzles have delighted people of all ages; designers are constantly at work inventing new and more difficult challenges. There are three-dimensional picture puzzles, double-sided puzzles, and puzzles with curving and irregular edges. A puzzle advertised as "the world's largest jigsaw puzzle" has 24,000 pieces. Monochromatic puzzles include "Little Red Riding Hood's Hood" (all red) and "Snow White Without the Seven Dwarfs" (all white).

Q What country hosts the world's oldest parliament?

A Contrary to popular belief, the world's oldest parliament is not in Britain. It's not in the United States, either.

First, a definition. A parliament is a representative assembly with the power to pass legislation and most commonly consists of two chambers, or houses, in which a majority is required to create and amend laws. (There are unicameral legislatures, however.) Congress became the supreme legislative body of the United States in 1789. The roots of the British Parliament date back to the 12th century, but it wasn't until 1689 that the Bill of Rights established Parliament's authority over the British monarch and gave it the responsibility of creating, amending, and repealing laws.

The title of Oldest Functioning Legislature in the World belongs to the Parliament of Iceland, known as Althing, which is Icelandic for "general assembly." Althing was established in A.D. 930 during the Viking age. The legislative assembly met at Thingvellir (about 30 miles outside of what is now the country's capital, Reykjavik) and heralded the start of the Icelandic Commonwealth, which lasted until 1262. Althing convened annually and served as both a court and a legislature. One of Althing's earliest pieces of legislation was to ban the Viking explorer Erik the Red from Iceland in 980 after he was found guilty of murder.

Even after Iceland lost its independence to Norway in 1262, Althing continued to hold sessions, albeit with reduced powers, until it was dissolved in 1799. In 1844, Althing was restored as an advisory body, and in 1874 it became a legislative body again, a function it maintains to this day. The parliament is now located in Reykjavik.

Considering the fact that it was created by a horde of bloodthirsty Vikings, Althing is an amazing testament to enduring democratic government.

Q Who told the first knock-knock joke?

A No one really knows for sure, but it couldn't have happened before the wooden door was invented, right? Could the "Knock, knock! Who's there?" phenomenon have started with the ancient Egyptians? According to historians, though, Egyptian genius was concentrated in the fields of mathematics, medicine, and architecture—not corny comedy.

As old and tired as knock-knock jokes may seem, they appear to be a fairly recent development. In August 1936, *Variety,* a trade magazine covering the entertainment industry, reported that a "knock knock craze" was sweeping America. Around the same time, British comedian Wee Georgie Wood debuted the catchphrase "knock, knock!" on his radio show. (It wasn't the setup to a joke, but a warning that a zinger was about to come.)

The knock-knock joke may have evolved from a Victorian party game called "knock, knock." According to language historian Joseph Twadell Shipley, the game started when a partygoer knocked on the door: "Who's there?" "Buff." "Buff who?" "Buff you!" The "buff" then tried to make the other guests laugh with wordplay or slapstick humor. ("Buff" in this sense is connected with "buffoon.") Whoever laughs first became the next buff, and the game began anew.

But the roots of the knock-knock joke may go deeper than this dreadful-sounding party game. In fact, some people credit William Shakespeare for inspiring the pun's classic pattern. How so? Dust off your old copy of *Macbeth* and turn to Act II, Scene 3. That's where you'll find Shakespeare's famous "porter scene," a satiric monologue delivered by a drunken porter who is pretending to be a doorman at the gates of hell:

> Knock, knock, knock. Who's there, i'th' name of Belzebub?—Here's a farmer, that hang'd himself on th' expectation of plenty: come in, time-pleaser...[knocking] Knock, knock. Who's there, i'th' other devil's name?—Faith, here's an equivocator, that could swear in both the scales against either scale; who committed treason enough for God's sake, yet could not equivocate to heaven: O! come in, equivocator. [knocking] Knock, knock, knock. Who's there?—Faith, here's an English tailor come hither for stealing out of a French hose: come in, tailor; here you may roast your goose.

Of course, today's knock-knock jokes aren't quite as clever or dramatic as Shakespeare's prose, but that may be their draw. Knock-knock jokes are simple wordplays that don't require a whole lot of thinking. And no matter how bad they might be—and they're usually really, really, really bad—they always get a reaction.

Orange you glad we didn't say banana? Sorry—we just couldn't resist that one.

Q Do you need a brush to paint the town red?

A You can bring a brush if you want to, but a wad of cash and a few friends will probably serve you better, because painting the town red means going out and having a wild time. The phrase is often used by people who have cause to celebrate: "Let's go out and paint the town red!"

How did painting become associated with raucous partying, and why red? One explanation dates back to England in 1837, when the Marquis of Waterford and some of his friends are said to have wreaked a little havoc in a town called Melton Mowbray; apparently, as part of their revelry, they painted some public buildings a crimson hue. Those wacky Brits! As fun as that sounds, it's unlikely that it accounts for the true origin of the phrase. The earliest published references are from the 1880s, and they all occurred in the United States.

One explanation posits that the saying may have come out of the nineteenth-century Wild West, where the unseemly parts of a town (i.e., the places where people had the most fun) were referred to as red-light districts. Lusty revelers who were ready for a night of no-holds-barred action may have vowed to carry their craziness anywhere they pleased, thereby making the entire town a red-light district. This explanation seems to involve more fun than the story of our friend the Marquis, but it calls for less actual paint.

The problem with identifying the origin of a phrase that is used to describe a night of drunken debauchery is that your average drunken debaucher tends to forget a lot of what happened during said debauchery. For this reason, we'll probably never know precisely how the term was coined.

So, no, you don't need a brush to paint the town red, but it's definitely a good idea to use a toothbrush the next morning.

Q Why is the area where prostitutes work called a red-light district?

A The first printed mention of a red-light district was made in 1894 in a Milwaukee newspaper, the *Sentinel,* but the semantic origin of the term is open to debate.

One explanation, which dates back to about the time of the *Sentinel's* article, relates to the red lanterns that railroad workers took with them into town when they had time off during a trip. The lanterns were kept lit so that the crew could be rounded up quickly if needed. If the men were engaging prostitutes, who tended to cluster in one section of town, that area was marked by many red lanterns. Others suggest that the term refers to the red shades that prostitutes of the early twentieth century put over candles and lamps as a discrete signal to passersby of their services.

The color red has long been connected to prostitution. Red paper lanterns were placed outside brothels in ancient China, and in the Biblical story of Rahab, a house of ill-repute was indicated with a red rope. During World War I, red lights in Belgian brothels indicated that the available services were for non-officers. (Brothels for officers sported blue lights.)

Color theorists—scholars who study the effects and meanings of colors to humans and animals across time and geographical boundaries—widely note that red is associated with sensual matters. In the United States, the color red has long been associated with the concepts of "stop" and "love." And aren't the women who work in a red-light district just trying to get men to stop and love?

Q Why is black the color of mourning?

A It's not clear whether black's negative connotations caused it to become associated with mourning, or if its link to mourning caused it to take on those connotations.

What is evident is that black's history as the color of mourning is a long one. The relationship goes back at least as far as the ancient Egyptians. The Roman Empire followed suit, and in later centuries the Roman Catholic Church's color sequence assigned black as a symbol of mourning. Indeed, this somber function feels natural. We seem to instinctively associate black with negativity—think of the dark feelings brought on by the passing of a loved one.

Black, however, isn't inexorably linked to doom and gloom in every society. Asian and some Slavic cultures consider white to be the color of mourning. In Buddhism, white is symbolic of old age and death. Brides in Japan wear white, but in contrast with Western culture, where bridal white is a symbol of purity or joy, a Japanese bride's white robe signifies her "death" from life at home with her parents. As you can see, when it comes to color associations, nothing is as simple as black and white.

WHY DO WE DO THE THINGS WE DO?

Q Why do we cross our fingers for luck?

A Humans are a superstitious bunch. We won't walk under ladders. We avoid black cats. We greet Friday the thirteenth with fear and trepidation. And when we need an extra jolt of good luck—when, say, we're confronted with a ladder, a black cat, or Friday the thirteenth—we cross our fingers. Why?

Tradition connects the gesture to the Christian sign of the cross. In earlier times, people saw supernatural evil just about everywhere they looked. They often attributed illness and misfortune to the influences of evil forces, and they expected to find spirits, witches, and other supernatural pests lurking around every corner. When they stumbled upon evil, it was common practice to call on divine protection by making the sign of the cross (touching the forehead, then the chest, then the left and right shoulders in turn).

It's possible that crossed fingers were originally meant as a similar appeal for God's protection. By subtly forming a cross with the index and middle fingers, a frightened Christian wouldn't attract undue attention. After all, you wouldn't want the neighborhood witch to know you were on to her—she might turn you into a mule or dispatch gophers to ravage your garden.

It's also possible that finger crossing predates the symbolism of the Christian cross. Some New Age spiritualists trace the gesture to a pagan practice in which two people would form a cross with each other's index fingers and then make a wish. In this case, the crossed fingers would have probably evoked the solar cross, an astrological sign that features an equilateral cross inside a ring and dates back to prehistoric times.

The solar cross was, among other things, associated with nature, earth, the sun, and divinity. According to this line of thinking, pagans believed that good spirits existed at the center of a perfect cross and that they could trap a wish there by making a cross with their fingers.

The practice of finger crossing continues to this day, even if we talk about it more than we actually do it. In fact, the phrase "keep your fingers crossed" dates back only to the 1920s. Even with the major strides we've made in understanding the universe around us, we remain, apparently, every bit as superstitious as our ancient ancestors were.

Q Why don't Scotsmen wear underwear beneath their kilts?

A Except for on a few formal occasions, no one tells these guys that they need underwear. It's their choice, and to the consternation and/or amusement of the rest of the Western world, many Scotsmen choose not to wear undershorts with their traditional kilts.

This harks back to the rough-and-tumble origins of the kilt—to the late sixteen hundreds or earlier, in the dark age before mankind had invented briefs, boxers, boxer-briefs, or Marky Mark, when the kilt was basically a huge bolt of wool that men wrapped around themselves in numerous ways, including as a shawl or a blanket.

Several Scottish military regiments stipulate that no underwear be worn unless there is a chance that things could get exhibitionist—like if you're marching in a pipe band (where knees are raised) or participating in the Highland games (where piping, drumming, dancing, and athletic contests take place). Not wearing underwear is called "military practice" or "going regimental," an idiom that is similar to the American expression "going commando," which refers to soldiers skipping the Skivvies in order to save rucksack space, stay cooler, avoid having to wash underwear, and prevent skin rashes and other nether issues.

Kilt-wearers—either civilians or soldiers—who do wear undies can buy them from makers of special underwear. These garments match the kilt patterns or are complementary in color. Another option is to wear boxers, briefs, or even Lycra shorts.

Again, the men in kilts usually make these vital decisions about their vitals. This unusual freedom only adds to the mystique of the Scottish soldier, who has been known throughout history as rugged to the point of uncouth—so secure in his manhood, in other words, that he doesn't care what goes on down there.

Q Why aren't all gas caps on the same side of the car?

A Your uncle's 1956 Chevy hid it behind the left taillight. On your dad's '65 Mustang, it was above the back license plate and beneath the galloping-horse emblem. Today's BMW has it on the right rear fender. Drive a Honda? Check the left rear fender. Own a Ford? It's on the right. Unless it's a Ford Fusion;

then it's on the left. We're talking gas caps, and they are no longer at the very back of the car because the fuel tank is no longer at the very back of the car. It's in front of the rear axle, protected from harm in a crash, and its filler neck enters from the side. But why isn't it on the same side from car to car?

Some speculate that it's always opposite the exhaust pipe to prevent fires from dribbled gas. Not true, say automotive designers; some cars, after all, have exhaust outlets on both sides. Another notion suggests that caps face the shoulder of the road so that luckless motorists needn't stand in traffic to refill a car that's run out of gas. But some cars from countries that drive on the right side of the road, like the United States, have left-side caps. And some cars from left-side-drive countries, like Japan and England, have right-side caps. A few optimistic souls even believe that automakers alternate sides so that we don't all queue up in the same line at the gas pump. They're mistaken.

For its part, the National Highway Traffic Safety Administration doesn't care which side the cap's on, as long as the car meets standards designed to reduce deaths and injuries from fires caused by fuel spillage during and after crashes. Tests set precise limits on the volume of fuel that can be spilled during rollovers and in rear impacts at specific rates of speed. How automakers satisfy the requirements is their own affair.

Once designers have a tank and filler neck that meet the standards, something they call "packaging" determines on which side the cap goes. Designers must route the filler neck around the suspension and exhaust components, trunk cavity, wheel housings, and door cutouts—all the stuff in the tail of a modern automobile.

Some manufacturers, like BMW and Honda, engineer all of their cars to standardize the cap on the same side. Others, like Ford and General Motors, follow the packaging dictates of each individual car. Some of today's fuel gauges display a little arrow telling us which side the cap is on. Sure, it's nice, but it's no galloping horse.

Q Why would one person try to get another's goat?

A The origins of this phrase are surprisingly literal. Indeed, there appears to have been a time in history when, for one reason or another, a person would break into a barn and abscond with another's goat. You can see how this would lead to the present-day usage of the phrase: Waking up in the morning to find yourself one goat poorer than you were the night before would surely be an aggravating experience.

So what's with all the goat thievery? There are a few possibilities. Anecdotal evidence suggests that goats were strategically placed in stables that housed racehorses. The presence of a goat was said to have a calming effect on horses, which would thereby perform better at the track. If the goat was removed, the horses became agitated. Racehorses don't perform as well in such a worked-up state, so gamblers looking for an edge might have tried to get a person's goat in order to affect the outcome of a race.

One source references an older belief that has to do with dairy cows. The goats served the same purpose—providing companionship in the stable. Cows were thought to produce more milk when a goat was hanging around. A rival farmer—or a regular troublemaker—might have tried to get a farmer's goat in order to stunt his cow's production.

A related American idiom, "Don't let them see where your goat is tied," gives some credence to these notions of the phrase's historical origins; if the thieves don't know where to find your goat, they'll have a hard time stealing it. Or, regarding the idiom's common usage, if a person doesn't know what irks you, he'll have a more difficult time making you lose your temper.

A less popular theory cites prison slang as the origin of the expression. Since 1904, at least, "goat" has been substituted for "anger." Thus, to get a person's goat would mean to incite anger. In this context, the phrase gives the same result, but involves fewer animals. The slang may have risen from the belief that goats are angry and ornery creatures, prone to lashing out and butting things with their horns.

Wherever the phrase originated—from gamblers, dairy farmers, or inmates—it persists. And attempting to find its exact source has gotten the goat of many a philologist over the years.

Q How did Xs and Os become shorthand for hugs and kisses?

A Are you one of those people who signs letters with cutesy Xs and Os to indicate hugs and kisses? Did you ever stop to wonder why you're doing that? What is so cuddly about an X or an O? Much like love itself, the how and why of it remain matters of conjecture.

The X can be traced as far back as the tenth century B.C., when it was used as the Paleo–Hebrew letter *Tav* and the symbol of the seal of Hashem (God), which stood for truth, completeness, and perfection. During the early Christian era, the X character signified the cross of Calvary (the Latin cross mounted on three steps) and was the first letter in Christ's name *(Xristos)*.

Fast-forward to the Middle Ages, when illiterate people supposedly substituted an X for their signatures. They would then kiss the mark, an act that was comparable to kissing a crucifix or Bible and implied a sworn oath. This practice continued until as recently as one hundred and fifty years ago.

According to the *Oxford English Dictionary,* the earliest known use of an X to signify a kiss was in 1763. However, this date is debatable—the practice may have started with handwritten notes, so documentation may be incomplete, which makes it difficult for lexicographers to pin down exact dates and sources.

For our purposes, the origins of the O are even more elusive. In his book *The Joys of Yiddish,* author Leo Rosten wrote that Jewish immigrants in the United States chose an O as their signature, as opposed to a cross symbol that represented Christ. Shopkeepers and salespeople also purportedly signed receipts using an O. Of course, this explanation does little to solve the hugs-and-kisses riddle.

There is even a running debate about which letter represents hugs and which represents kisses. If you look closely, you will see that the outline of an X can suggest a silhouette of the union of two pairs of lips. Dear Uncle Ezra, an online question-and-answer forum from Cornell University, posits that the X resembles a single pair of lips pursing for a kiss. Some researchers contend that the X implies a hug because it resembles two pairs of crisscrossing arms. On the other hand, the O could be the hug, as it might represent a pair of arms encircling another person. Or it could be the kiss—use your imagination and you'll see the pouty imprint of a smooch.

Whatever the particulars, Xs and Os should continue to flourish in this era of shorthand text messaging. Unless they are elbowed out by LOL ("lots of love"), of course. To that unthinkable possibility, we say, OMG ("oh my god").

Q Why do they drive on the left side of the road in England?

A Or if you're from England: Why do Americans drive on the right side of the road? At any rate, it probably goes back to the good old Romans, who, we're fairly certain, drove on the left. Bryn Walters, a British archaeologist, discovered a Roman track in England that led to an ancient quarry. On the left-hand side, running out of the quarry, the track was deeply grooved by wagon wheels; Walters surmised that this was because the wagons were laden with stone and, therefore, were heavier. How did he reach this conclusion? The other side of the road was less deeply worn, presumably because it was used by empty wagons that were going into the quarry to collect their loads.

Fast-forward to the Middle Ages, when you never knew who you were going to meet on the road (though you could be pretty sure that he'd be smelly). A popular "left side" theory is that people needed to be able to easily whip out their swords to defend themselves. Because most people were right-handed, they rode on the left to leave their sword arms free to whack foes.

Regardless of how things went down, the British rode on the left long before the "keep left" law was introduced in the seventeen hundreds. So why have so many other countries opted for the right? One explanation posits that when Napoleon started conquering countries in the early eighteen hundreds, he forced the vanquished to drive on the right, in defiance of the hated English.

America, meanwhile, liked the thought of being revolutionary—and the idea of annoying the English—so it also went for the right. According to another explanation, modes of transportation developed differently in different places. In the United States and France, a typical horse-drawn wagon had no driver's seat. Instead, the driver sat on the left-rear horse so that his right arm was free to wield a whip and persuade his good-for-nothing beasts to go faster. In order to communicate with other riders, the driver would ride on the right-hand side of the road so he was next to any oncoming traffic.

Britain's horse-drawn carts had driver's seats. Again, the driver would typically use his right hand to whip the horses. Since he didn't want the whip to get caught up in the load behind him as he swung, he sat on the right. Again, he needed to keep an eye on, and communicate with, oncoming traffic, so he drove on the left.

About two-thirds of the world's countries now drive on the right; among those still on the left are Britain, India, Australia, South Africa, and Japan. So if you're terrified of trying to drive a car on the left and you happen to be in one of those countries, a word of advice: Use public transportation.

Q Are good manners the same around the world?

A Sit up straight. Say please and thank you. Don't put your elbows on the table. Most of us were drilled from an early age in proper manners and etiquette. But once you leave your home country, things get a bit complicated. Here are some examples of how cultures do things differently around the world.

1. In China, Taiwan, and much of the Far East, belching is considered a compliment to the chef and a sign that you have eaten well and enjoyed your meal.

2. In most of the Middle and Far East, it is considered an insult to point your feet (particularly the soles) at another person, or to display them in any way, for example, by resting with your feet up.

3. In most Asian countries, a business card is seen as an extension of the person it represents; therefore, to disrespect a card—by folding it, writing on it, or just shoving it into your pocket without looking at it—is to disrespect the person who gave it to you.

4. Nowadays, a bone-crushing handshake is seen as admirable in the United States and UK, but in much of the East, particularly the Philippines, it is seen as a sign of aggression—just as if you gave any other part of a person's body a hard squeeze!

5. Orthodox Jews will not shake hands with someone of the opposite sex, while a strict Muslim woman will not shake hands with a man, although, to confuse matters, a Muslim man will shake hands with a non-Muslim woman. People in these cultures generally avoid touching people of the opposite sex who are not family members.

6. When dining in China, never force yourself to clear your plate out of politeness—it would be very bad manners for your host not to keep refilling it. Instead, you should leave some food on your plate at each course as an acknowledgment of your host's generosity.

7. In Japan and Korea, a tip is considered an insult, rather than a compliment, and, for them, accepting tips is akin to begging. However, this tradition is beginning to change as more Westerners bring their customs with them to these countries.

8. The "okay" sign (thumb and forefinger touching to make a circle) is very far from okay in much of the world. In Germany and most of South America, it is an insult, similar to giving someone the finger in the United States, while in Turkey it is a derogatory gesture used to imply that someone is homosexual.

9. Similarly, in the UK, when the two-fingered "V for victory" or "peace" salute is given with the hand turned so that the palm faces

inward, it is considered extremely rude, having a meaning similar to raising the middle finger to someone in the United States.

10. In Greece, any signal that involves showing your open palm is extremely offensive. Such gestures include waving, as well as making a "stop" sign. If you do wish to wave goodbye to someone in Greece, you need to do so with your palm facing in, like a beauty pageant contestant or a member of the royal family.

11. In many countries, particularly in Asia and South America, it is essential to remove your shoes when entering someone's home, while in most of Europe it is polite to ask your host whether they would prefer you to do so. The reason, as anyone who's ever owned white carpet will attest, is simple hygiene and cleanliness.

12. Chewing gum might be good for dental hygiene, but in many parts of the world, particularly Luxembourg, Switzerland, and France, public gum-chewing is considered vulgar, while in Singapore most types of gum have been illegal since 1992 when residents grew tired of scraping the sticky stuff off their sidewalks.

Q What's behind the tradition of flying flags at half-mast?

A As you might have guessed, the custom of flying a flag only midway up its pole has nautical roots. The convention of lowering the colors to half-mast to symbolize mourning probably started in the fifteenth or sixteenth century, though no one knows precisely when. Nowadays, the gesture is recognized almost everywhere in the world.

The first historical mention of lowering a flag to recognize someone's death comes from the British Board of the Admiralty. In 1612, the British ship *Hearts Ease* searched for the elusive

Northwest Passage, a sea route through the Arctic Ocean that connects the Atlantic to the Pacific. During the voyage, shipmaster James Hall was killed. When the *Hearts Ease* sailed away to rejoin its sister ship, and again when it returned to London, its flag was lowered to trail over the stern as a sign of mourning.

That all who saw the *Hearts Ease* understood what the lowered flag meant suggests it was a common practice before then. Starting in 1660, ships of England's Royal Navy lowered their flags to half-mast each January 30, the anniversary of King Charles I's execution in 1649.

In the United States, the flag is to be flown at half-mast (or half-staff) on five designated days: Armed Forces Day (the third Saturday in May), Peace Officers Memorial Day (May 15), until noon on Memorial Day (the last Monday in May), Patriot Day (September 11), and Pearl Harbor Remembrance Day (December 7). In addition, according to the United States Code, the flag goes to half-mast for thirty days following the death of a U.S. president, past or present, and for ten days following the death of the sitting vice president, a current or retired chief justice of the Supreme Court, or the speaker of the House.

But it doesn't stop there. For justices of the Supreme Court other than the chief justice, as well as for governors, former vice presidents, or cabinet secretaries of executive or military departments, the flag is lowered until the person is buried. For a member of Congress, the flag flies at half-mast for the day of and the day after the passing. By presidential order, the flag can also be lowered for the deaths of "principal figures" of the government or foreign dignitaries, such as the pope.

Q Whatever happened to dunce caps?

A Ah, dunce caps. Those tall, conical paper hats shamed many a struggling student back when our classrooms were a little less enlightened than they are today. It would be a stretch to say that dunce caps represent a proud tradition, but they certainly are part of a long-standing one.

Surprisingly, dunce caps date back hundreds of years. Even more surprisingly, they were named after a real guy. And that unfortunate guy's name was John Duns Scotus. (Duns was his family name, while Scotus was a Latin nickname meaning, roughly, "You know, that guy from Scotland.") He was a philosopher, Franciscan friar, and teacher who lived during the late Middle Ages. He is still remembered as the founder of a dense and subtle school of philosophy called Scotism. He was influential during his lifetime, and today he is regarded as one of the most important philosophers of his era.

His arcane and convoluted logic seemed like the height of sophistication at the time, and it inspired a school of followers—known as the Dunsmen or, more casually, Dunces—who emulated his academic style and dominated the universities of Europe. But by the 16th century, a new intellectual movement was attacking the old traditions. The Renaissance humanists hated the obscure and overly complicated method of reasoning that the Dunces employed. They labeled the Dunsmen as "old barking curs" who lacked the ability to reason, and began using "dunce" as an insult to describe a thickheaded person.

And this is where the headgear comes in. One of Scotus's stranger opinions was particularly easy for the humanists to ridicule: He had claimed that conical hats actually make you smarter by funneling knowledge down to your brain. (This also explains why cones were the hat of choice for wizards, by the way.) After the humanists succeeded in turning "dunce" into an insult, the cone hat became the official headwear of the stupid.

How these peculiar hats made it into schools isn't entirely clear, but by the 19th century, American and European teachers punished ignorance by making students wear paper dunce caps and sit in the corner of the classroom. The idea was to encourage kids to learn by shaming them when they didn't.

These days, dunce caps occasionally pop up in cartoons, but they're no longer standard classroom equipment. Dunce caps went out of vogue at around the same time as corporal punishment, and for the same basic reason. Beginning in the 1950s, B. F. Skinner and other behaviorist psychologists demonstrated that positive reinforcement—rewarding desired behavior—is a far more effective way to motivate students than punishment. According to Skinner, people "work harder and learn more quickly when rewarded for doing something right than when punished for doing something wrong," and he maintained that punishment should be a last resort in the classroom.

Skinner's beliefs slowly took hold. By the 1980s, enough Americans disapproved of harsh punishment that spanking and shaming became rarities in public schools. While some teachers and parents still swear by the power of the paddle, nobody seems to feel strongly enough about the dunce cap to defend it as a learning tool. Ol' John Duns Scotus can finally rest in peace.

Q What's the point of the barber pole?

A Barbers date back to Egypt in the Bronze Age (circa 3500 B.C.). At the dawn of their profession, they were medicine men and priests. Barbers were placed at the head of their tribes and treated with utmost respect—and with good reason. A popular superstition of the time was that good and evil spirits entered and left the body through the hair. The only way to be rid of an evil spirit was to get a trim. In addition to tonsorial exorcisms, barbers baptized babies and arranged marriages.

By the Middle Ages, the barber's duties had been somewhat refined, though they were still rather broad by today's standards. On top of the usual shaves and haircuts, barbers performed minor medical procedures, such as bloodletting and pulling teeth. Bloodletting is the process of using leeches to drain blood from the body, and it was prescribed for such diverse conditions as fevers, inflammations, and even hemorrhages. The practice continued until the late nineteenth century.

The design and placement of the barber pole were inspired by bloodletting. The original wooden pole was gripped tightly by the patient, which caused his or her veins to bulge and made them easier to puncture. The poles are thought to have been painted red to mask bloodstains. Once the procedure was completed, the pole was placed outside and the linens that had been used to clean up the mess were hung on it to dry. The linens flapped in the wind and became twisted around the pole, which created the red-and-white design that is so familiar today.

The pole, with its linens flapping in the wind, served as an advertisement for the bloodletting business. The metal top of the pole, which has evolved into a spherical shape, was originally meant to signify the bowl in which the leeches were kept before a bloodletting; the bottom represented the bowl into which the leeches were placed once they were bloated with blood.

Few people today realize that the barber once held such an esteemed position in society. In contemporary society, the barbershop has been reduced to a scribble on a list of errands, somewhere to go between the hardware store and the grocery store. You might have a friendly chat with your barber while your hair is being cut, but an exorcism? That's better left to religious professionals.

Q Why are flowers placed on graves?

A This tradition can be traced to the ancient Greeks, who performed rites over graves that were called *Zoai.* Flowers were placed on the graves of Greek warriors; it was believed that if the flowers took root and blossomed on the graves, the souls of the warriors were sending a message that they had found happiness in the next world.

The ancient Romans also used flowers to honor soldiers who died in battle. The Romans held an elaborate eight-day festival during February called *Parentalia* ("Day of the Fathers"), during which roses and violets were placed on the graves of fallen soldiers by friends and family members.

According to acclaimed historian Jay Winik, the tradition began in America at the end of the Civil War, after a train had delivered Abraham Lincoln to his final resting place in Springfield, Illinois. In his Civil War book *April 1865,* Winik writes: "Searching for some way to express their grief, countless Americans gravitated to bouquets of flowers: lilies, lilacs, roses, and orange blossoms, anything which was in bloom across the land. Thus was born a new American tradition: laying flowers at a funeral."

Following Lincoln's burial, people all over the country began decorating the graves of the more than six hundred thousand soldiers who had been killed—especially in the South, where organized women's groups also placed banners on the graves of soldiers. The practice became so widespread that in 1868, General John Alexander Logan—the leader of the Grand Army of the Republic, a Union veterans' group—issued an order designating May 30 as a day for "strewing with flowers or otherwise decorating the graves of comrades who died in defense of their country." The day was originally called Decoration Day, but it later became Memorial Day. On May 30, 1868, thousands gathered at Arlington National Cemetery in Virginia to decorate more than twenty thousand graves of Civil War soldiers. In 1873, New York became the first state to declare Decoration Day a legal holiday.

Today, the tradition is stronger than ever. In addition to being placed on graves, flowers are often displayed in funeral homes and churches for burial services. The most elaborate arrangements are positioned around the casket, perhaps hearkening back to the belief of the ancient Greeks that a flower in bloom signifies happiness in the afterlife.

Q Why do people shake hands?

A In today's Western world, the handshake serves a number of purposes. It can be a greeting or a farewell, and it can signify congratulations or condolences. In business, a deal can be sealed with a handshake.

Where exactly did this quirky little custom originate? This question is the object of much debate, conjecture, and confusion, mainly because the handshake likely predates written history. Some historians trace the origin of the modern Western handshake—the clasping of hands—to medieval Europe. Back then, shaking hands was hardly a congenial gesture—it was more like a shakedown. Men would clasp hands to make sure that neither was concealing a weapon.

Eventually, the handshake matured into something more civilized. According to Philip A. Busterson's 1978 book *Social Rituals of the British,* writer and explorer Sir Walter Raleigh, who was noted for his manners, introduced the handshake that we know today in the late sixteenth century.

Since then, the handshake has become a nuanced custom. For instance, it's considered poor taste to greet someone with a handshake that's too strong; a limp handshake, on the other hand, is perceived as a sign of weakness. It is viewed as an insult to refuse a handshake or to fake a greeting by offering but not following through on a handshake.

Today, there are high-fives, low-fives, soul shakes, secret handshakes—you name it. The handshake's grip on civilization has tightened considerably since those days in medieval Europe, when some skin on skin meant, "I'm not going to kill you…at least not today."

Why is a white flag a symbol of surrender?

It seems like a cliché straight out of a 1950s B-movie or an episode of *Hogan's Heroes.* Despondent and fearing for their lives, the vanquished search desperately for anything white—a handkerchief, a shirt, a pair of underpants—and attach it to a stick. They then proceed cautiously (or, in the case of *Hogan's Heroes,* clumsily) toward their gloating foe.

In reality, the tradition of the white flag as a symbol of surrender or truce goes back a couple of thousand years. In the West, the Roman historian Tacitus mentioned a white flag of surrender that was used at the Second Battle of Cremona in A.D. 69. In the East, the use of a white surrender flag is believed to date back just as far.

It's unclear how the color white first came to symbolize surrender. Flag experts surmise that it happened because white is a neutral hue, one that could be easily distinguished from the colorful banners that armies often carried into battle. Today, the use of the white flag as a sign of peace or surrender is an official part of the rules of warfare, as referenced in the Geneva Conventions.

The white flag has had other military uses throughout history, though none lasted long. For a short time during the Civil War, the Confederacy used a mostly white national flag that was known as the "Stainless Banner"—however, it caused confusion in battle and was scrapped. During the sixteen hundreds, the French (those lovable contrarians) used a white flag to signify the intent to go to battle. Historians don't tell us whether the French looked with disdain at anyone who didn't understand their unconventional use of the white flag—but we can guess that they did.

Was a day always twenty-four hours long?

From almost the first moment humans decided that a day needed more segmentation than the obvious day and night, twenty-four has been the number of choice. The practice began thirty-five hundred years ago, with the invention of the Egyptian sundial.

Before then, humans had little interest in specifying the time of day. It was enough to differentiate between morning and evening by using an obelisk—a four-sided monument that cast a westward shadow in the pre-noon hours and an eastward shadow as evening approached.

The sundial divided a day into twelve hours of light and twelve hours of darkness—though, of course, it was capable of displaying the time only during sunlight hours. The Egyptians marked an hour of twilight at sunrise, an hour of twilight at sunset, and ten hours in between. Nighttime hours were approximated by the use of decan stars (stars that rise in the hours before dawn). Twelve decan stars rise in the Egyptian sky during the summer months.

Historians believe that these stars might be the reason the earliest timekeepers chose to base their sundials on the number twelve. The hour did not have a set length until the second century B.C.—thirteen hundred years later. Until then, the length of the hour changed as the seasons changed. During winter months, nighttime hours were longer; during summer months, daytime hours were longer. In the second century B.C., the Greek astronomer Hipparchus divided the day into twenty-four equal segments.

However, keeping time was still an inexact science for the common man. Until the invention of the modern clock, just about everyone simply divided a day into twelve hours of daylight and twelve of darkness. The modern mechanical clock, which keeps time by the regulated swinging of a pendulum, was conceptualized by Galileo, the sixteenth- and seventeenth-century mathematician and philosopher. The first such timepiece was built by Dutch scientist Christiaan Huygens in the seventeenth century.

In the centuries since, clocks have become increasingly precise, culminating with invention of the atomic clock in the twentieth century. Finally, the entire population of the world is in sync. Hipparchus would be proud.

Q Why do people knock on wood for good luck?

A Finally, things seem to be going right. When you comment on that fact, however, you have an irresistible urge to add the phrase "knock on wood" and frantically seek out the nearest table, desk, chair, or doorjamb so that you can rap your knuckles on it a few times for good luck. But why is it good luck?

Many people point to the rosary as the origin of the "knock on wood" phenomenon. Rosaries, or "prayer beads," are a series of wooden beads strung together, often with some sort of religious icon, usually a wooden crucifix, where the two ends join. In times of prayer, Christians rub the rosary beads. The tradition dates back to at least the fourth century, when Christians flocked to Constantinople to touch the cross on which Jesus was believed to have been crucified—it was thought that this artifact would bestow blessings and healing upon those who touched it. Pilgrims would touch the cross three times, in deference to the Christian Holy Trinity; this practice eventually evolved into a series of shorter, faster motions, leading to the "knocking" of today.

Another religious theory comes from the pagan notion of nature spirits. By knocking on wood, one is thought to invoke the protection of the spirits that reside within the wood. Some even go so far as to knock only on the underside of a table or other piece of furniture, believing that to do otherwise is to bop the spirit on the head, which is not very polite and could serve to anger the spirit.

And then there's the mundane suggestion that knocking on wood for good luck stems from a simple game of tag. *The Boy's Own Book,* published in England in 1829, lays out the rules of this now-ubiquitous playground game: A wooden (or iron, in some versions) post or a tree is designated as a "safe spot," and anyone who touches it is immune from being tagged. Some of us never grew out of that notion, and to this day, when things go south, we seek the "safety" of nearby wooden objects.

However it got here, by 1908 the phrase had entered American lexicon. An article in the *Indianapolis Star* that year used the phrase at least twice, in its familiar, modern form. The article focused on a promising young athlete, and the author took care to add the parenthetical phrase "knock on wood!" after assertions of the athlete's potential prowess. No word on whether these embellishments helped propel the athlete to greatness.

Q Why is it considered bad luck to walk under a ladder?

A It's not just bad luck to walk under a ladder—it's an all-around bad practice. Even the most ardent skeptics, those who pay no mind to a black cat or a newly cracked mirror, recognize the practical danger inherent in ducking under a ladder. Three words can sum it up: open paint can. Who knows what might be precariously balanced atop the highest step (the "Not a Step" step)? One wrong move and you might find yourself showered with Cinnamon Speckle Semi-Gloss.

But what about the superstitious aversion to ladders? This goes back at least as far as the sixteen hundreds. In this period, the theory goes, when the condemned were being led to the gallows to be hanged, they were forced to walk under the ladder that led to the last platform their boots would ever be planted on; the executioner, meanwhile, walked around it. This ritual, witnessed weekly by an enthralled mass of people, brought about the association of everyday ladders with death in general and the gallows in particular. For the average person, walking under a ladder was the same as tempting fate, invoking his or her own death.

Another theory has the triangular shape formed by a ladder leaning against a building representing the Christian Trinity—the Father, Son, and Holy Spirit. Breaking a triangle, which was thought to be sacred, was considered blasphemous; defying God in such a way was tantamount to declaring an allegiance with the Devil. People avoided walking under ladders to keep their friends and neighbors from thinking they had made such a pact. In times when the popular vote could get a man hanged for alleged involvement in witchcraft and other devilry, quashing such rumors before they started might have been a matter of life and death.

At any rate, skeptics and believers alike should be cautious when confronted with a ladder. Whether the outcome is an eternity in Hell or an afternoon washing latex paint out of your hair, taking a few extra steps around a ladder seems like a small price to pay for personal safety.

Q Why do people say "bless you" when someone sneezes?

A A sneeze—*sternutation* is the medical term for it—is acknowledged in many different languages. In Arabic, for example, a polite person says *alhamdu lillah* ("praise be to God") when someone sneezes. If you speak Telugu, you'll say *chiranjeeva,* which means "live for an eternity." In German, *gesundheit* is the usual response to an "achoo!" It's *jai mati ji* in Hindi, and *na zdrowie* in Polish. Yet there isn't a routine comment to reply to belching, farting, coughing, snoring, or having one's joints creak. Why sneezing?

If you think about it, sneezing is an odd reflex. It even appears to border on the perilous. Anthropologists guess that prehistoric man might well have acknowledged a sneeze as a means of giving thanks that the person had been given another chance at life. The sneezer's breath, and very essence, had abruptly exploded from the body, but the breath immediately after the sneeze restored life, something to be thankful for.

What about modern society, specifically the English-speaking segment? Why do people say "bless you" after someone sneezes? There are many theories, and a prominent one has its roots in Rome. (Doesn't everything?) In AD 590, Pope Gregory I asked for unending prayers by the people and for chanting priests to parade through the streets in hopes of warding off an epidemic of bubonic plague. Sneezing was considered an early symptom of the disease, and the "bless you" was meant to ask for the blessings of God.

That might not have been a bad idea even if modern theories of disease transmission hadn't yet developed—a sneeze can spray thousands of bacteria-filled droplets. Which is why we customarily cover our mouths when we sneeze.

Sneezing isn't dangerous. Some people even find it pleasurable enough to use snuff, or ground tobacco, to induce sneezing.

By the way, the urban myth that your eyes will pop out of your head if you sneeze with your eyes open is just that—an urban myth. Scientists aren't sure why people reflexively close their eyes when they sneeze, but there's no evidence that your eyes will pop out if you hold them open. So sneeze away—someone undoubtedly will be waiting to wish you well.

Q Why are coffins six feet under?

A We've all heard the line in a cheesy movie that, against our better judgment, has sucked us in and stuck us to the couch: "One more move, and I'll put you six feet under." Whether the words are growled by a cowboy in a black hat or a mobster in pinstripes, everyone knows what those six feet represent: the depth where coffins reside after burial.

Or do they?

The bad guy may well mean what he says, but the final resting place for someone unfortunate enough to be in a coffin varies depending on the site of the funeral. Burial depths can range from eighteen inches to twelve feet. There's no world council that has decreed that a person must be put to rest exactly six feet under. Think about it. Digging a six-foot grave in a region below sea level, such as New Orleans, would get pretty soggy. (Of course, a corpse floating up from the grave might add some much-needed flair to that cheesy movie.)

Most grave depths are determined by local, state, or national governments. New Orleans has dealt with its topographical issues by placing most of its dead above ground in crypts. The area's gravesites in the ground are almost always less than two feet deep—and even that doesn't prevent the occasional floater.

The California requirement is a mere eighteen inches. In Quebec, Canada, the law states that coffins "shall be deposited in a grave and covered with at least one meter of earth" (a little more than three feet). This is similar to New South Wales in Australia, which calls for nine hundred millimeters (slightly less than three feet). And the Institute of Cemetery and Crematorium Management in London says that "no body shall be buried in such a manner that any part of the coffin is less than three feet below the level of any ground adjoining the grave."

If burial depths vary from place to place, how did the phrase "six feet under" come to life? (Sorry, couldn't resist.) Historians believe it originated in England. London's Great Plague of 1665 killed seventy-five thousand to one hundred thousand people. In Daniel Defoe's book *A Journal of the Plague Year,* he writes that the city's lord mayor issued an edict that all graves had to be dug six feet deep to limit the spread of the plague outbreak. Other sources confirm Defoe's claim.

Of course, the plague is a scourge of the past, and today's world has no uniform burial depth. But who really cares? It still makes for a winning line in an otherwise schlocky movie.

BACK IN THE DAY

Q Where did the idea for the weekend come from?

A According to the Book of Genesis, the answer is simple: The idea came from God, who rested on the seventh day after creating the universe during the previous six. But the weekend as we know it—mounting giddiness after lunch on Friday; two days of sports, yard work, and beer; and zombified Monday-morning staring—is a relatively recent development.

The ancient Jews introduced the concept of non-work at the end of the week. Their time was structured around observing the Sabbath, or Shabbat, as a day of rest and worship. In ancient Rome, the seven-day week was officially adopted in A.D. 321, and the early Christians chose Sunday as their Sabbath.

The weekend as a time for chilling out—with worship optional—started taking shape in Britain during its industrial revolution. In the early eighteen hundreds, large numbers of people moved to cities and into factory jobs; they logged a six-day work week, and Sunday was their official day off. In this new economy, regular people could earn enough to enjoy leisure time. That led to partying on Saturday evening and through most of Sunday. In the mid-eighteen hundreds, workers began taking Monday off occasionally to, among other things, sleep off their Sunday drunkenness.

Since so many people were skipping work on Monday, and it was taboo to stage public events on the Christian Sabbath, Monday became a day for sporting events, festivals, and performances. Religious groups weren't happy that Sunday had become a day of general debauchery. Neither were factory owners, who couldn't count on sufficient numbers of workers showing up on Monday.

So, in the 1870s, many factories turned Saturday into a half day. In effect, management struck a deal with workers: We'll reduce the workload so that we can count on you showing up. Religious groups were pleased, too: Workers could get more partying out of their systems on Saturday, which tipped Sunday back into the worship category. The *Oxford English Dictionary* traces the first known description of this day and a half as a "week-end" to an 1879 issue of the English magazine *Notes and Queries.*

"Week-ends" also caught on in the United States, and in 1908, a New England spinning mill changed Saturday to a full holiday so that Jewish workers could observe Shabbat. The custom spread and gained a foothold in 1926 when Henry Ford adopted it at his automobile plants. Ford wasn't just being a nice guy: More leisure time meant more travel, which meant more auto sales.

The Great Depression solidified the two-day weekend: A five-day work week divided the available jobs among more people and, thus, reduced unemployment. If only they could have squeezed an extra day of leisure in there.

Q What did they use for birth control in the old days?

A Tough question. It's not information that got written down—at least not often.

Some hints are found in the ancient writings of doctors and naturalists. They show that fruits and herbs played a big role in controlling fertility. An Egyptian scroll of medical advice that's thirty-five hundred years old tells how to end a pregnancy at any point: Mix the unripe fruit of acacia with honey and other items to be soaked up by an absorbent pad of plant fiber. Insert the pad into the vagina, and an abortion will follow. (Dissolved acacia produces lactic acid, which is a spermicide.)

Other old texts show that herbal birth control was brewed into teas. The leaves of pennyroyal (a type of mint) and parts of many plants that look like weeds to us could be brewed as a "morning-after" cure for unwanted pregnancies. Juniper berries, willow bark, mugwort, aloe, anise, dittany, and certain ferns were all used. Seeds from the plant Queen Anne's lace or from pomegranates were eaten for the same reason, as were figs.

Many of these plant-based solutions were known to people in the Middle Ages and the Renaissance. In Shakespeare's *Hamlet,* Ophelia plays with the herb rue, a weed known to induce abortions. Rue is found throughout the Americas, too, and many native groups used it to end pregnancies.

What about condoms, sponges, and other devices? Illustrations that are thousands of years old show men using condoms, though the earliest condom that's been found dates to 1640. It's made of sheep intestines. Goodyear began mass production of rubber condoms in 1843.

Female condoms, or cervical dams, have also been around for millennia. They've included seedpods, oiled paper, seaweed, lemon or pomegranate halves, beeswax, and even moss. History also records the use of spermicidal potions made from oils, vinegar, rock salt, wine, and herbs.

Superstition and ignorance about the properties of certain ingredients played roles in some of these birth-control methods. Modern research on animals shows that some of these ingredients would have resulted in lower pregnancy rates or increased instances of miscarriage. However, some of these substances also are toxic, which would have made the birth-control benefits moot.

Q What was the Year Without a Summer?

A It lives in infamy, particularly in New England and Europe. The year 1816 is also remembered as the Year With No Summer, the Poverty Year, and Eighteen Hundred and Froze to Death.

The Year Without a Summer came toward the end of a period that climatologists call the Little Ice Age. This was a span of three hundred to nearly seven hundred years (the science community doesn't universally agree on when the Little Ice Age started). Regardless of its length, temperatures the world over averaged about one degree Celsius lower than normal during its duration. One degree might not seem like much, but the change set the stage for some unpredictable weather patterns.

What caused the Little Ice Age is not completely known. Scientists have identified two potential contributors working in concert: intense worldwide volcanic activity and decreased solar activity.

Scientists now generally attribute the strange weather in 1816 to the 1815 eruption of Tambora, an Indonesian volcano. Volcanic activity caused unusual weather at other points in history, particularly in 1991, when an eruption in the Philippines ushered in a less-drastic period of global cooling. Volcanic eruptions can send huge amounts of ash into the atmosphere, filtering out the sun's rays. The eruption of Tambora sent up eight times more volcanic material than the 1991 eruption did; hence, the cataclysmic effects. The regions of New England and Europe had some of the most drastic weather abnormalities; these are also the areas that yield the most recorded data relating to early-nineteenth-century weather.

New England had snowfall at high elevations and killing frosts every month in 1816. Brief periods of respite from the cold, which gave farmers hope that they would be able to harvest normal crops, lasted no more than a couple of weeks, and then the temperatures plummeted again. Some warm periods were even shorter: June 5, for instance, saw temperatures in the mid- and upper seventies across New England; just two days later, towns in Massachusetts reported near-freezing temperatures. Newspapers reported weather more severe than during an average winter storm; the Danville, Vermont, North Star, in what can surely be called an understatement, called the period "gloomy and tedious."

Short-term effects in the area included rising agricultural stock prices. Many crops were reduced severely, some to half or a quarter of their usual yield, leading to a fear of widespread famine. Many settlers were so disheartened by these events that they packed their bags and traded New England for the western frontier.

The effects of Tambora's eruption were similar in Europe. Farmers went bankrupt; food prices skyrocketed; hungry citizens rioted and confiscated edible imports for themselves. Switzerland declared a national emergency and released instructions as to which wild plants were edible and which were not. Famine and poor health in 1816 led to a typhus epidemic, which killed more than one hundred thousand people across Europe.

Weather records from the era from other parts of the world are spotty, so the full effect of Tambora's eruption isn't fully understood. It is known that most of Asia saw a delayed monsoon season that was more violent than usual, and that these poor conditions might have contributed to the cholera epidemic that ravaged Asia for years before it spread to Europe and parts of Africa. The cholera epidemic was particularly devastating for Europe, which was still recovering from the typhus epidemic at the time.

While 1816 is remembered as the Year Without a Summer in the Western world, climate studies have shown that some regions experienced unnaturally high temperatures. In the Middle East, for example, temperatures were above average. Weather patterns returned to "normal" the following year, though it would be another thirty years before the Little Ice Age came to an end.

Q Who got to pick the Seven Wonders of the World?

A Humans love their lists—to-do, grocery, pros and cons—so it makes sense that we would obsessively list cool stuff to see. There are "wonder lists" for every kind of wonder imaginable.

The most famous is the Seven Wonders of the Ancient World, or "Six You'll Have to Take Our Word for, Plus One You Can Actually See." One of the earliest references to "wonders" is in the writings of Greek historian Herodotus from the fifth century B.C. Herodotus wrote extensively about some of the impressive wonders he had seen or heard about. However, the concept of the Seven Wonders didn't really catch on until the second century B.C.

For the next fifteen hundred or so years, six of the seven were a lock, often appearing on compiled lists, with the seventh spot being a rotating roster of hopefuls. By the time the list became the accepted seven of today (around the Renaissance), the Lighthouse of Alexandria had taken up the seventh spot.

No one person actually got to pick the seven; it was more a generally accepted concept based on the frequency with which certain wonders landed on different lists. Furthermore, by the time the Middle Ages rolled around, most of the wonders couldn't be seen in their full glory because of damage or destruction, so the selections were based primarily on reputation. The seven wonders were the Pyramids of Giza in Egypt, the Hanging Gardens of Babylon in Iraq, the Statue of Zeus at Olympia in Greece, the Mausoleum of Maussollos at Halicarnassus in Turkey, the Colossus of Rhodes, the Temple of Artemis at Ephesus, and the Lighthouse of Alexandria.

The Pyramids of Giza are the Energizer Bunny of the wonders. They were the oldest when the lists began, and they are the only wonder still standing. The Colossus of Rhodes was big in size (107 feet high) but not big on longevity—the statue stood for only fifty-four years but was impressive enough to stay in the minds of list-makers.

In 2001, the New 7 Wonders Foundation was established by a Swiss businessman. The foundation's intention was to create a new list of seven wonders of the world based on an online vote (notably, anyone could vote more than once). In 2007, the wonders chosen were Chichén Itzá in Mexico, Christ the Redeemer in Brazil, the

Colosseum in Rome, the Great Wall of China, Machu Picchu in Peru, Petra in Jordan, and the Taj Mahal in India. The Great Pyramids of Giza were given honorary finalist status after Egypt protested that these great historical structures shouldn't have to compete against such young whippersnappers.

So make your own lists, and bury them in a time capsule. Maybe one day that cool treehouse you built with your dad will finally get its due.

Q If the Maya were so brilliant, why aren't they still around?

A They are! Six million people who are living in Mexico, Honduras, Guatemala, and Belize speak Mayan or identify themselves as Maya. The Nobel Peace Prize winner in 1992, Rigoberta Menchú, was Maya.

But you're thinking of the ancient Maya, right? The temple-building, bloodthirsty warriors seen in the movie *Apocolypto*? The great astronomers and priest-kings who ruled over the sculpted plazas deep in the jungle? Those Maya? Well, today's Maya are the descendants of those Maya—the Mayan people who flourished in southern Mexico and the Yucatan Peninsula for millennia.

Mayan civilization was established well before 1500 B.C. For centuries, the Maya cut and burned jungle land to plant crops, and after 400 B.C., they began building cities, temples, astronomical observatories, and pyramids. Samples of their complex writing date back to before the time of Christ. Their accomplishments peaked between A.D. 300 and A.D. 800, when the Maya engineered reservoirs, canals, and irrigation systems to maximize the food supply that supported their growing network of city-states. It was a remarkable civilization.

Then, abruptly, the Mayan people abandoned their cities, and the great buildings and plazas fell into ruin. Signs of violence and fire mark some (though only some) sites. The population in A.D. 900 was merely a third of what it had been one hundred years earlier. Why? Invasion? Disease? Political upheaval?

Recent scientific studies show that a series of multiyear droughts began in A.D. 750 and lasted for nearly three hundred years. No one knows for sure, but the drought could have caused famine, epidemics, invasions, wars, riots, and rebellion. Many Maya died, and others left their cities and took to the jungle, where they tried to survive.

Thanks to GPS and satellite imaging, more Mayan sites and ruins have been spotted in recent years. Digs and explorations are constantly evolving. Mayan writing remained a mystery until the 1980s, when it was finally deciphered. So the answer to the riddle of what drove the Maya from their cities could yet be found.

Q Why did they put coins in the eyes of dead people?

A As if death wasn't a big enough downer, it accounts for one of the biggest expenses the average person will ever incur. From caskets to flowers to embalming services, people pay through the nose for a decent burial. And until fairly recently, you could say that the dead also paid through the eyes.

The most common explanation for the tradition of putting coins on the eyes of the dead points to the mythical Greek figure of Charon, one of the Western world's first undertakers. Typically depicted as a morose, somewhat creepy old man (undertakers haven't changed much in three thousand years, apparently), Charon was a ferryman who conveyed the souls of the dead to the land of the dead across the rivers Styx and Acheron.

Much like today's undertakers, Charon's services came at a price. The cost of a ride to Hades was an *obol,* an ancient Greek coin valued at one-sixth of a drachma (the equivalent of fifteen to twenty dollars in today's world). Placing the coins on the corpse's eyes, the Charon theory holds, was to ensure that the soul would reach its final destination.

Though this explanation seems convincing, a thorough review of Greek mythology identifies one major flaw in the argument: Charon, who was a picky sort, would only accept payment if the coin was placed in the mouth of the corpse. Hence, the tradition in ancient Greece was to put coins in dead people's mouths, not in their eye sockets.

However, the ancient Greeks were not the only ones who believed that money was necessary in the afterlife. Many civilizations, including the Egyptians and the Incas, buried their dead with money and other treasure to make sure that the deceased were comfortable in eternity. So it is possible that the erroneous attribution to Greek mythology is rooted in fact.

A more likely explanation, though, is a combination of this mythology and simple practicality. When a human body dies, rigor mortis sets in. One of the more troubling aspects of this condition is the tendency for the eyelids to pop open. Not only is it creepy to have a dead person staring at you, but in many ancient traditions, it was considered bad luck for the dead person to cross into the afterlife with his or her eyes open—the corpse, it was believed, would look for someone to take with it. Coins—which fit nicely into the eye sockets, are weighty enough to counteract the effects of rigor mortis, and could be used as emergency funds at a post-mortem tollbooth—were the ideal solution.

Among some people, the custom of putting coins in the eyes of the dead continued until at least the late nineteenth century, when mortuary scientists figured out how to deal with rigor mortis. Besides, undertakers no longer require payment from the orifices of the dead. Just those of the living.

Q Why did the ancient Romans begin their year in March?

A The Romans claimed that Rome's first king, Romulus, came up with the first calendar and that he decided the year would begin on the spring equinox. Most years, this falls on the day we call March 20. Since Rome was supposedly founded in 735 B.C., that became year one of the Roman calendar.

We can only guess why the spring equinox was chosen. Maybe it had meaning because the world comes to life again after a cold winter: Flowers bloom, greenery appears, and birds build nests. If that is the reason, it's worth it to note that no European cultures began the year with spring. Some of the ancient Greeks began their year with the summer solstice (June 21); the Celts picked November 1 as New Year's Day; and the Germanic tribes started their year in the dead of winter, much as we do today. Bottom line: We don't know why the Roman year started in springtime.

The original name of March was Martius, which was an homage to the god of war, Mars. Romulus designated only ten months for the year, though. Why? Romulus liked the number ten. He organized his administration, his senate, his land, and his military legions into units of ten, so why not his calendar, too?

However, ten times thirty or thirty-one (the designated numbers of days of the months back then) made for a pretty short year.

Records don't survive to tell us how the people of Rome managed, but within a couple of generations, two more months were added to the calendar.

Did the year continue to start on the spring equinox? Not exactly. Maintaining the calendar was the duty of priestly officials, who could add days when needed. Over the centuries, corrupt priests and politicians manipulated the Roman calendar to extend political terms of office and delay important votes in the assembly—they didn't really give a hoot if it ended up astronomically accurate. The first month of every year was March, but it didn't always correlate to the March that we know—it was sometimes as many as three months off.

Julius Caesar—who was once one of those priestly officials—revised the calendar when he took control of Rome. He brought it more in line with the calendar that we know today; in fact, he even added a leap year. But Caesar's leap year was a little different from ours: Once every four years, February 24 was counted twice. Those wacky Romans.

Q: Why did so many Europeans move overseas between 1850 and 1914?

A: There are almost as many reasons as there were immigrants. We'll list the big ones.

The poor and downtrodden often believed that they had to leave their home countries or die. Ireland endured years of famine, and more than a million people fled the island. Groups of people in Germany, Czechoslovakia, and Hungary suffered political retaliation after a series of failed revolutions. Parts of Scandinavia were economically depressed. The Amish and Mennonites were pushed from Eastern Europe by religious persecution, while pogroms drove two million Russian Jews from their villages. In many places, land and jobs were scarce, which left families on the brink of starvation.

These desperate people saw their salvations across the ocean. News of discoveries of gold and a vast, open frontier attracted them to America. Often, one or two men would venture overseas; usually, they would describe success and easy money in letters to friends and family. Cousins and neighbors were quick to follow their acquaintances across the ocean. By 1900, 94 percent of those who made the move to the United States had relatives or good friends waiting for them.

In some parts of Europe, men had always left home to seek seasonal jobs. Migrant workers traveled to major cities such as London and Vienna to work in factories, then returned to the countryside for harvest season. Once railroads connected more cities and steamship travel became affordable in the 1870s, those men could travel farther to find better jobs. America, with its myriad opportunities, was within reach. The majority of these immigrants were men; a great number worked in the U.S. for years, then returned permanently to Europe. Since they left no descendants in America, they are largely forgotten.

Hundreds of thousands of others stayed in the United States and raised families. In the mid-nineteenth century, most of these immigrants came from Western Europe. By the 1890s, Eastern Europeans (Hungarians, Greeks, Russians, and Poles), as well as Sicilians, were flocking to America. At the same time, the migration of Britons and Germans had already peaked and was in decline.

Immigration slowed to a trickle when World War I started in 1914. In 1917, the United States established a literacy requirement in order to block immigration by those who could not read. Further restrictions—such as a quota system that limited the number of people who could immigrate to the U.S. from each country—came in the 1920s.

Since then, regulations have varied but have not been lifted entirely. The United States has never gone back to the wide-open system of the nineteenth century.

What was the first toothbrush?

Well, it was more of a stick than a brush. A chew stick, to be exact.

Men and women have used tools to get gunk off their teeth since the dawn of civilization in ancient Mesopotamia (modern-day Iraq). There, in the cosmopolitan city of Babylon, your typical well-groomed urbanite would find a nice, solid stick and chew on one end until it was frayed and softened. This well-gummed twig was perfect for excavating stray bits of food that were stuck in dental crevices. And if it failed to do the trick, there was always the *siwak*—a narrow, sharp implement made from a porcupine quill or a long thorn (ouch!). In other words, it was a glorified toothpick.

After the Babylonians, the Egyptians adopted the siwak, and since then the toothpick has never gone out of fashion. In Renaissance Europe, nobles probed their gums with toothpicks that were crafted from precious metals—how's that for bling?—and even today, individually wrapped toothpicks are available at many restaurants.

So when did bristles—the defining feature of the contemporary toothbrush—first come into play? Their earliest use was probably in China. By the fifteenth century, the Chinese were using wild boar's hairs for bristles, attaching them to bamboo or bone handles. However, this great leap forward in dental care didn't make its way to Europe, where the bristled toothbrush was developed independently, many years later.

In the seventeenth century, Europeans cleaned their teeth with little rags or sponges that they dipped into a solution of either salt or sulphur oil. But William Addis of Clerkenwald, England, had a better idea: In 1780, he gathered and trimmed hair from cows' tails for bristles and then fastened the hair with wire into small holes that were bored into a handle made of cattle bone. It was—drum roll, please—a toothbrush. The invention caught on across Europe, often with bristles made of boar's hair (and, more rarely, horsehair).

The toothbrush continued to be made in much the same way until the twentieth century. Food shortages during World War I led to the confiscation of all cattle bones so that soup could be made from them.

This triggered the next step in the evolution of the toothbrush: Handles were crafted from celluloid, the first plastic. Japan's invasion of China in the late 1930s caused another shortage—this time, of the boar's hair that was used for bristles.

In 1938, DuPont de Nemours unveiled its new miracle fiber, nylon, and was soon manufacturing toothbrushes with nylon bristles. But would you believe that 10 percent of today's toothbrushes are still made with boar's hair bristles? So much for progress.

Q Who invented cigarettes?

A A bunch of seventeenth-century beggars, bless their enterprising souls.

In 1614, King Philip III of Spain established Seville as the tobacco capital of the world when he mandated that all tobacco grown in the Spanish New World be shipped there to control its flow and prevent a glut. Seville specialized in cigars, but beggars found they could cobble together cigar scraps, wrap them in paper, and make passable cigarettes, called *papeletes* ("little papers").

You can trace the growth of the cigarette in Britain and America to the cultural ramifications of wars that wracked Europe between the French Revolution in the late eighteenth century and the Crimean War in the mid-nineteenth century. During the French Revolution, the French masses made a social statement by smoking cigaritos. Produced from the tobacco that was scared up from leftover snuff, cigars, and pipes, these cigaritos were unlike the aristocracy's snuff. In the mid-eighteen hundreds, the cigarette was brought to Britain by soldiers who had returned from the Crimean War, where they had learned of cigarettes from their French and Turkish allies.

Cigarettes began to rise in popularity in the United States during the Civil War. Soldiers received tobacco in their rations and enjoyed rolling their own smokes with the sweet tobacco that was grown in the Southeast. (The first cigarette tax was imposed during the Civil War.) By the late eighteen hundreds, cigarettes were being hand-rolled in factories in England, Russia, Germany, and the United States.

In 1880, the industry was revolutionized by the invention of the cigarette-rolling machine. This device not only could produce many times the number of cigarettes as a human could roll by hand, but it also could do so more cheaply. James Bonsack, a Virginian, invented the machine, which created a long tube of paper-wrapped tobacco that was cut into cigarette lengths.

A few years after the machine was invented, tobacco industrialist James Duke licensed it and worked out the bugs. Less than a decade later, Duke was manufacturing four million cigarettes a day. Accompanying increased production was the introduction of a more easily and deeply inhaled variety of tobacco, as well as plenty of advertising.

About forty years after Bonsack's invention, cigarette production had increased roughly thirtyfold, leaving the previously popular cigars in the dust. And at the turn of this century, an estimated 5.5 trillion smokes were manufactured annually worldwide. Hey, got a light?

Q Whatever happened to Neanderthals?

A Once upon a time, about one hundred thousand years ago, there were people who lived in the mountains of Europe. Their bodies were short and stocky, and they had barrel chests, bowed legs, and sloping shoulders. Their faces were characterized by thick protruding foreheads, big noses, and receding chins. They used tools made of bone, stone, and wood; wore clothing consisting of animal hide; and cooked with fire. When one of them died, the body was interred in a ceremony and sometimes strewn with flowers. They may have even played music with flutes that were fashioned from hollow bones.

Then, about forty thousand years ago, some very interesting neighbors showed up. They were smaller and slimmer, and sported longer legs and narrower fingers. They also had a more pronounced jaw, which made it easier for them to articulate a variety of sounds. Among their innovations were language, jewelry, art, and tools with sharp, finely honed blades. After another ten thousand years, the first inhabitants had disappeared. But the later arrivals flourished. If you want to see one of their descendants, just look in the mirror. That's you—*Homo sapiens.*

And what happened to *Homo neanderthalensis?* Neanderthals and *Homo sapiens* share a common ancestor, *Homo erectus,* who evolved in Africa about two million years ago. For decades, paleontologists wondered if the two groups had been biologically close enough to interbreed. In other words, scientists theorized that Neanderthals didn't actually die out—they suspected that they're still with us, in our genes. In 2006, however, biologists at the Max Planck Institute for Evolutionary Anthropology in Leipzig, Germany, and the Joint Genome Institute in Walnut Creek, California, retrieved DNA from a fragment of a thirty-eight-thousand-year-old Neanderthal femur bone and concluded that it was highly unlikely that Neanderthals and *Homo sapiens* produced mutual offspring. Though related, they were probably two distinct species, which would have made interbreeding impossible.

So what did happen? One dark scenario casts *Homo sapiens* as war-like aggressors who attacked and killed the peaceable Neanderthals. Dramatic as this theory is, researchers consider it as unlikely as Neanderthal-sapiens love children.

Another possible culprit is one that we're worrying about today: climate change. The disappearance of the Neanderthals coincided with the end of the last ice age. Receding glaciers altered the landscape and affected animal migration patterns. Perhaps

Neanderthals found survival difficult in this warmer world. Disease, too, may have played a role in their extinction.

In the end, no one really knows why Neanderthals died out. But our interest in their demise has led us to uncover a wealth of information about their lives. Like us, Neanderthals had big brains. They lived in social groups and performed rituals, just as we do today.

Contemporary humans are the only species of Homo left on the planet, and while we may glory in our singularity, being one of a kind can be a little lonely. This may be why our imaginations are drawn so powerfully to the ancient campsites of these distant relations, who were lost forever just as our own history began.

Q What if a duel ended in a tie?

A Ah, the romantic days of yore, when courtly ladies rode in horse-drawn carriages and gentlemen who knew the meaning of honor slapped each other with gloves and then met at dawn to fire guns at each other. To quote poet/rocker Ray Davies, "Where have all the good times gone?"

Back in the good old days, if a gentleman felt insulted, he didn't stoop to starting a shouting match, a fistfight, or even a flame war on an Internet discussion board. Instead, he had recourse to a duel. Dueling, which originated in sixteenth-century Italy before gaining popularity throughout Europe, generally followed a protocol. The insulted party would throw down his glove before demanding "satisfaction" from the other party. Apparently, this satisfaction could only be obtained by shooting at him.

What happened if a duel ended in a tie? Who won satisfaction?

Actually, duels that ended in a tie were rather commonplace. Not all duels were fought "to the death"; duels could also be fought "to the blood," in which the first man who drew blood from his opponent was the victor. In the case of pistol dueling, "to the blood" meant that each man was allowed only one shot. (Ultimately, the act of the duel itself was enough to save the honor of the participants, and many a duel ended with the two parties simply firing into the air.)

Pistol dueling is what most people think of when they imagine duels, though before the advent of guns, weapons such as swords were used. And even after pistols became fashionable, gentlemen sometimes chose other weapons to defend their honor. One apocryphal tale tells of an aborted duel in the mid-eighteen hundreds between Otto von Bismarck and his nemesis Rudolf Virchow, in which sausages (those wacky Germans!) were chosen as the weapon.

One of the most famous duels that ended in a tie occurred in 1826 between two United States senators, Henry Clay of Mississippi and John Randolph of Virginia. Clay was known as a firebrand, and when he and Randolph disagreed on an issue, Clay demanded satisfaction. The much-ballyhooed duel took place on April 8. Each senator was allowed one shot; naturally, they both missed. It could be seen as an example of the inability of politicians to do much of anything right.

Why did the Romans sell urine?

Because there was a demand for it. Why the demand? Because Romans used the stuff by the bucketful to clean and dye clothing. Why urine? Because it worked, it was plentiful, and it was cheap. Why on earth did it work? Because the nitrogenous urea in urine generates ammonia when the urine is left standing around, and this ammonia acts as both a disinfecting and bleaching agent.

Some Romans, like many other people of the time, used urine to wash their teeth, too. But before you go dissing the Romans, realize that for more than fifteen hundred years after the Roman Empire peaked, Europeans were still using urine to clean clothes.

And the Romans were not slovenly people, relatively speaking. They were quite the scientists. For example, it's been argued that after the fall of Rome, battlefield medicine didn't return to Roman levels until World War I, and that's partly a function of hygiene. Besides, lots of people today drink their own urine in the name of alternative medicine.

The Romans made extensive use of public baths—a bit of a turn-off to many of us today, but actually a sign of their culture's advancement. (The Romans were great innovators in matters hydraulic, as evidenced by their clever work with aqueducts and plumbing.) In the first century A.D., the emperor Vespasian enacted a "urine tax," and with it coined the proverb *pecunia non olet* ("money does not smell"). But pee does. Imagine the troughs at the more than one hundred public baths where urine vendors would collect their wares, which they sold to the multitude of establishments around Rome and elsewhere that cleaned and bleached and performed a kind of dry-cleaning on woolens. A significant number of Romans were employed in the cleaning industry, experts say.

All in all, we moderns would be astonished to learn how "green" the ancients were. They didn't pump crude oil from the earth and make gasoline of it, and they didn't make plastics of whatever it is we use to make plastics. No, they used what was at hand in remarkable ways. And considering how much urine is quite literally at hand, it's no surprise that they found a way to use it.

Q Why did ancient Egyptians shave their eyebrows?

A Shaving away all bodily hair, including eyebrows, was part of an elaborate daily purification ritual that was practiced by Pharaoh and his priests. The ancient Egyptians believed that everything in their lives—health, good crops, victory, prosperity—depended on keeping their gods happy, so one of Pharaoh's duties was to enter a shrine and approach a special statue of a god three times a day, every day. Each time he visited the shrine, Pharaoh washed the statue, anointed it with oil, and dressed it in fresh linen.

Because Pharaoh was a busy guy, high-ranking priests often performed this duty for him. But whether it was Pharaoh or a priest doing it, the person had to bathe himself and shave his eyebrows beforehand.

Shaving the eyebrows was also a sign of mourning, even among commoners. The Greek historian Herodotus, who traveled and wrote in the fifth century BC, claimed that everyone in an ancient Egyptian household would shave his or her eyebrows following the natural death of a pet cat. For dogs, he reported, the household members would shave their heads and all of their body hair as well.

Herodotus was known to repeat some wild stories in his books—for instance, he reported that serpents with bat-like wings flew from Arabia into Egypt and were killed in large numbers by ibises. Herodotus claimed he actually saw heaps of these serpent skeletons. So you might want to take his eyebrow-shaving claim with a grain of salt…and a pinch of catnip.

Q Why do old churches have steeples?

A Because they are pointing to heaven. Other reasons have been offered over the years, but clergymen and historians generally agree that steeples atop churches are meant to guide a person's gaze skyward.

Religious buildings have led the eyes heavenward for millennia. Egyptians had their obelisks, for example. It could be argued that these structures are phallic symbols, but the practical fact is that towers and pinnacles make temples and other religious buildings easy to see. And they fill believers with awe.

As far back as the Dark Ages, watchtowers were features of churches, which were often the biggest buildings in town. Documentation is hard to come by, but at some point the towers began to serve less as perches for watchmen and more as cubbies from which to hang bells and as mounts for crosses that could be seen for miles. Architects began adding purely decorative spires to Christian churches in the twelfth century, when Gothic architecture was all the rage.

The wooden steeple as we know it today came into vogue later. On September 2, 1666, a fire destroyed much of London. Thirteen thousand homes were incinerated, along with more than eighty churches. King Charles II commissioned Christopher Wren, considered one of England's greatest architects, to rebuild St. Paul's Cathedral and about fifty other churches. Wren topped one of his first projects, St. Mary-le-Bow, with a steeple, and Londoners were duly inspired. The city was soon filled with steeple-topped churches, and colonists carried the architectural style to America.

Steeples are no longer church staples everywhere in the United States—the custom is disappearing in California and other western states. In the South, however, most congregations wouldn't think of building a church without a steeple. A steeple continues to be excellent housing for a church bell—and these days, it is just as likely to be a hiding place for microwave antennas for cell phones.

Q Who was Mansa Musa?

A History's epic gold rushes were generally characterized by masses of people trekking to the gold. But in 1324, the legendary Mansa Musa bucked the trend by trekking masses of gold to the people—and severely depressed the Egyptian gold market as a result.

If you could talk to a gold trader from 14th-century Cairo, he might say that the worst time of his life occurred the day Mansa Musa came to town. Musa, king of the powerful Mali Empire, stopped over in Cairo during his pilgrimage to Mecca. Arriving in the Egyptian metropolis in 1324, Musa and his entourage of 60,000 hangers-on were anything but inconspicuous. Even more conspicuous was the 4,000-pound hoard of gold that Musa hauled with him.

While in Cairo, Musa embarked on a spending and gift-giving spree unseen since the pharaohs. By the time he was finished, Musa had distributed so much gold around Cairo that its value plummeted in Egypt. It would be more than a decade before the price of gold recovered from the Mali king's extravagance.

Musa ruled Mali from 1312 until 1337, and ushered in the empire's golden age. He extended Mali's power across sub-Saharan Africa from the Atlantic coast to western Sudan. Mali gained tremendous wealth by controlling the trans-Sahara trade routes, which passed through Timbuktu and made the ancient city the nexus of northwest African commerce. During Musa's reign, Mali exploited the Taghaza salt deposits to the north and the rich Wangara gold mines to the south, producing half the world's gold.

Musa's crowning achievement was the transformation of Timbuktu into one of Islam's great centers of culture and education. A patron of the arts and learning, Musa brought Arab scholars from Mecca to help build libraries, mosques, and universities throughout Mali. Timbuktu became a gathering place for Muslim writers, artists, and scholars from Africa and the Middle East. The great Sankore mosque and university built by Musa remain the city's focal point today.

Musa's story is seldom told without mention of his legendary pilgrimage, or *hajj,* to Mecca.

The *hajj* is an obligation every Muslim is required to undertake at least once in their life. For the devout Musa, his *hajj* would be more than just a fulfillment of that obligation. It would also be a great coming-out party for the Mali king.

Accompanying Musa was a flamboyant caravan of courtiers and subjects dressed in fine Persian silk, including 12,000 personal servants. And then there was all that gold. A train of 80 camels carried 300 pounds of gold each. Five hundred servants carried four-pound solid-gold staffs.

Along the way, Musa handed out golden alms to the needy in deference to one of the pillars of Islam. Wherever the caravan halted on a Friday, Musa left gold to pay for the construction of a mosque. And don't forget his Cairo stopover. By the time he left Mecca, the gold was all gone.

But one doesn't dish out two tons of gold without being noticed. Word of Musa's wealth and generosity spread like wildfire. He became a revered figure in the Muslim world and inspired Europeans to seek golden kingdoms on the Dark Continent.

Musa's journey put the Mali Empire on the map—literally. European cartographers began placing it on maps in 1339. A 1375 map pinpointed Mali with a depiction of a black African king wearing a gold crown and holding a golden scepter in his left hand and a large gold nugget aloft in his right.

ALL IN THE NAME

Q Do Dutch couples always split the bill?

A Ah, to be living in Amsterdam. Legalized marijuana? Check. Legalized prostitution? Check. Wooden shoes? Check again. And as if this weren't enough, conventional wisdom holds that dating is considerably cheaper in Holland—at least if the phrases "going Dutch" and "Dutch treat" have any validity.

Although some Americans today might think of the Netherlands (the official name for Holland) as a country of tulips and debauchery, this wasn't always the case. For a brief period in the seventeenth century, Holland was one of the world's most powerful empires, largely due to its early exploitation of spice-producing lands in Asia and the Pacific. Along with its financial and military might, Holland saw a cultural flowering during this period, with painters like Rembrandt and Vermeer churning out masterpieces and scientists like Christiaan Huygens laying the foundation for the theory of light. (We're not artists, but we're told this was important.)

Of course, Holland wasn't the only imperial nation in the seventeenth century. Most of the other European countries were also getting busy plundering and looting the rest of the world. England, one of the biggest offenders, didn't like the fact that Holland was horning in on its territory. The British expressed their displeasure by waging not one, not two, but three wars against the Netherlands during the seventeenth century. Unfortunately for the British, they were forced to an unsatisfying draw in the first and were soundly whipped in the next two.

Unable to defeat the Dutch in battle, they did the next best thing: They made fun of them. During the seventeenth century, a series of phrases deriding the Dutch worked their way into the English language, such as "Dutch concert" (pandemonium), "Dutch courage" (alcohol), "Dutch comfort" (no comfort at all), and "Dutch feast" (when the host of a dinner gets hammered before the guests even arrive). Most of these phrases have gone the way of the Dutch flotilla, but one has made its way through the centuries to our modern lexicon: "going Dutch" or "Dutch treat," meaning that everyone pays his or her own way.

"Dutch treat" regained popularity in the United States in the late nineteenth century, when xenophobic Americans spewed invectives at German immigrants whom they mistakenly referred to as Dutch (a mispronunciation of the German word *deutsch,* which translates to "German"). Back then, it was considered cheap and rude to make somebody pay his or her own share for an outing that you suggested. Today, however, "going Dutch" is standard in most situations, and it's becoming increasingly so in dating etiquette.

Besides, there isn't any real reason to make fun of the Dutch anymore. By the eighteenth century, Holland's military power had waned and the country slowly receded from the world stage. Holland remains notable, however, for its progressive social policies, such as decriminalized marijuana, same-sex marriages, and socialized health care. Though we're not sure we want to know what "Dutch medicine" entails.

Q Is Chinese food from China?

A Well, that's a dumb question. Of course Chinese food is from China. But what we're getting at is whether the modern idea of Chinese takeout is less East and more West. In that context, Chinese food is definitely a product of the United States.

In nearly every case, so-called "American Chinese" foods were inspired by counterparts from China. Not surprisingly, the American versions of Chinese foods are more meat-based and less dependent on vegetables than dishes that originate in the Far East. General Tso's chicken, sesame chicken, Chinese chicken salad, chop suey, chow mein, crab rangoon, fried rice, and Mongolian beef are among the many items at Chinese restaurants that are essentially American derivatives of staples from the motherland.

The origin of nearly every menu item at a typical American Chinese establishment is hotly contested. However, perhaps the most popular staple of these restaurants, the fortune cookie, is indisputably American—or more specifically, was created by a Chinese immigrant in the United States.

As with any great invention, several parties have been credited with thinking up the fortune cookie, including a Japanese landscape architect named Makota Hagiwara, who some say distributed the treat in San Francisco in the early nineteen hundreds. However, it is widely thought that Los Angeles baker and Chinese immigrant David Jung, later the founder of the Hong

Kong Noodle Company, first handed out cookies containing encouraging words (the fortunes) to homeless Californians in 1918. Since Jung was a Presbyterian minister, the strips of paper he inserted in his cookies featured Bible scripture.

By the 1930s, several fortune-cookie factories were in production. The paper-filled treats were folded by hand and inserted using chopsticks until 1964. Today, fortune cookies are a hit everywhere. Even in…well, China. Fortune cookies first began surfacing in Asia simply because American tourists asked for them.

Q Why does Swiss cheese have holes?

A Rumors continue to run rampant about this age-old question. Some say manufacturers allow mice to nibble on Swiss before packaging the cheese. Others insist crafty deli owners cut the holes by hand with their carving knives. However, both of these conspiracy theories have more holes than, well, Swiss cheese.

Truth be told, and it's a bit embarrassing, Swiss cheese has holes because it has bad gas. That's right. Those holes in your sweet, nutty Swiss are actually popped bubbles of carbon dioxide gas.

Where do these gassy bubbles come from? Well, all cheese begins with a combination of milk and starter bacteria. The type of bacteria used helps determine the flavor, aroma, and texture of the finished cheese product. In the case of Swiss, cheese-makers use a special strain of bacteria called *Propionibacter shermani.*

During the curing process, when the cheese ripens, this *P. shermani* eats away at the lactic acid in the cheese curd, tooting carbon dioxide gas all the while. Swiss cheese is a densely-packed variety with a thick, heavy rind, so this built-up gas has nowhere to go. Trapped inside, the gas forms into bubbles. These bubbles eventually pop, leaving behind the characteristic holey air pockets.

In formal cheese lingo, these holes are referred to as "eyes." And the art of cheese making is such that their sizes can be controlled. By adjusting acidity, temperature, and curing time, dairies can create a mild baby Lorraine Swiss with lacy-looking pinholes or a more assertive Emmentaler Swiss with eyes the size of walnuts.

Oddly, in the United States, the size of Swiss cheese holes is subject to United States Department of Agriculture regulation. Every wheel of Grade A Swiss that is sold in America must have holes with diameters between three-eighths and thirteen-sixteenths of an inch.

All of this goes to show that sometimes, it's best not to overthink your cheese. Just slap it on a cracker, pour a glass of wine, and enjoy.

Q When the French swear, do they say, "Pardon my English"?

A They probably should, but they don't. The phrase "Pardon my French" has an elusive origin, but it likely grew out of the long-standing rivalry between England and France. As a result of their history of mutual contempt, each country's everyday language contains many stock phrases and terms that denigrate the other.

The French, for example, have long been thought of in the English mind as champions of indecency and lewdness. The terms "French pox" and "French disease" were used by the English to describe syphilis and other venereal diseases, beginning as early as the sixteenth century. And we only have to ponder the images evoked by phrases like "frenching," "French kiss," and "French tickler" to get a glimpse of France's reputation. (Okay, you can stop pondering the images now.)

But the French weren't about to take this lying down. One of their more inventive phrases was *les Anglais ont débarqué,* which translates to "the English have landed." Fair enough, until you learn that they used it to describe menstruation. This phrase probably stemmed from the bright red uniforms of the English soldiers who flooded into France to fight against Napoleon. The English were associated with an unwelcome crimson arrival, and this morphed into a euphemism for menstruation.

In the nineteenth century, both countries came up with similar terms for things, simply swapping "French" and "English" as appropriate. A "French letter," for example, was an English euphemism for a condom, while a Frenchman would have preferred a *capote anglaise* (an "English hood"). "To take French leave" means "to leave without saying goodbye"; *filer à l'anglais* means "to flee like the English."

There isn't a similar symmetry with the phrase "Pardon my French." When the French swear and decide to apologize—after all, many people swear constantly without feeling sorry about it—they generally say, *Excusez moi* ("excuse me"), or they use another faintly regretful phrase. Such a response is logical, but hardly insulting. And what fun is that?

What is the difference between a city mile and a country mile?

In some countries, the mile is a standard unit of measure for a distance that equals 5,280 feet. Why it's 5,280 feet can be traced to the ancient Britons. Freed from Roman rule, the Britons decided upon a compromise between the Roman *mille passus* (a thousand paces, which was five thousand Roman feet) and their furlong (660 British feet, which naturally differed slightly in length from a Roman foot). The statute mile (now called the international mile) is 5,280 feet, or eight furlongs.

If that weren't confusing enough, there are different kinds of miles: the nautical mile, the geographical mile, the air mile, the metric mile. One you won't usually see, however, is the country mile. That's because it has no numeric definition. The *Merriam-Webster Online Dictionary* dates the term to about 1950, and most dictionaries say that "country mile" is informal and used to denote "a long distance."

Does this mean that miles are longer in the country than elsewhere? Although the origin of the phrase is unknown, there are several theories about its meaning. One holds that in times during which people walked nearly everywhere, a mile didn't seem very far. In the era of cars and mass transportation associated with city living, walking a mile is a bigger deal.

Another possibility relates to the grid systems of streets in many cities versus the meandering roads often found in the country. The shortest distance between two points is a straight line, but rural roads sometimes are anything but straight. So it can in fact take you longer to get from one place to another in the country, where two points that are a mile apart may not be connected by a straight road.

So if you have to go a country mile, you should be prepared for a long journey.

Q Why are Poland natives called Poles, but Holland natives aren't called Holes?

A Have you ever heard the phrase *pars pro toto?* Don't worry if you haven't—it's not exactly something that would come up during your dinner conversation. It's Latin for "a part for the whole." When talking geography, this refers to the practice of using a small part of a country to describe the whole country (or a larger area). For instance, many people refer to the United Kingdom as England, when in reality the United Kingdom consists of England, Scotland, Wales, and Northern Ireland.

Why are we mentioning this string of words in a dead language? Because Holland is a perfect example of *pars pro toto.* When people talk about Holland, they're most often referring to the country called The Netherlands. Holland is a western part of The Netherlands that consists of two provinces, North Holland and South Holland. It's kind of like if you were to call the United States New England or the Southwest.

That's all well and good, but it still doesn't explain why we don't call people from Holland "Holes," so let's explain it now. First off, Holland does not sound the same as Poland. Poland is pronounced Pole-and, but Holland is not pronounced Hole-and. So it's a bit lazy to use "Hole" as a shorthand.

But "it sounds wrong" isn't the only reason "Hole" isn't used. See, Pole comes from a Polish word that means "field dwellers," and there's no comparable description for Holes. Not only that, English-speaking people often refer to folks from The Netherlands as Dutch. The word "Dutch" first popped up in the fourteenth century, and it came from the German word for the Germanic people, *Deutsch.* By the sixteenth century, Dutch referred to anyone who spoke a Germanic language. (English people in the

Middle Ages weren't the most sensitive bunch.) Over the years, Dutch and German eventually came to mean what they do today. So to many English speakers, anyone from Holland is simply Dutch. People from The Netherlands, however, usually refer to themselves as *Nederlanders.*

It's interesting to note that the name Holland, rather than The Netherlands, is often used to promote the country because it's the term that's more recognized around the world. Holland is also used by some Dutch people informally, though there are areas in which calling someone a Hollander is an insult. Oh, and they don't all wear wooden clogs and pick flowers in front of windmills. But that would be lovely, wouldn't it?

Q Is the North Star always in the north?

A The answer is yes, but it's not quite that simple. Although the North Star is always in the north, it isn't always the same star. The North Star is also called the Pole Star because it is the star most directly above Earth's North Pole. It appears due north of the observer, and the angle between it and the horizon tells the latitude of the observer. Consequently, the North Star has been used for navigation for thousands of years.

However, the North Star has limited capability as a navigational tool. Because it is only visible in the Northern Hemisphere, it is of no help when you travel south of the equator. There is no precise Southern Hemisphere equivalent to the North Star, although the constellation Crux, or the Southern Cross, points towards the South Pole.

But back to the North Star and why its identity changes: Due to the precession of the equinoxes (which is a fancy astronomer's term for "the Earth wobbles when it turns"), the axis upon which our planet rotates shifts ever so slightly. As the shift occurs over many centuries, another star in the distance elbows its way in as the useful North Star. Currently, it's Polaris, a middling bright star at the end of the Little Dipper's handle, about 430 light years from Earth.

Time is running out for Polaris, just as it did for its predecessors. In 3000 B.C., the star Thuban in the constellation Draco served as the North Star. In A.D. 3000, Gamma Cephei, or Alrai, will get the call. Iota Cephei will have its turn in A.D. 5200, followed by Vega in A.D. 14000.

For now, though, Polaris will guide you if you are lost. Unless you're someone from then Northern hemisphere who happens to be adrift off the coast of, say, Peru. Then you're pretty much screwed.

Q Did the French really invent French toast?

A Not so fast, mon ami. There are plenty of conflicting stories about which country was the first to dunk day-old bread into milk and fry it.

The accounts go all the way back to ancient Rome. Take a look at *Apicius de re Coquinaria*—a collection of Roman cookery recipes that was compiled in the fourth or fifth century AD and translated to English in 1936 by professional chef Joseph Dommers Vehling—and you will find a dish titled *Aliter Dulcia* ("Another Sweet").

The recipe for *Aliter Dulcia* goes as follows: "Break [slice] fine white bread, crust removed, into rather large pieces which soak in milk [and beaten eggs]. Fry in oil, cover with honey, and serve."

Sounds similar to the golden-brown slices topped with maple syrup that you find at IHOP today, right? Still, no one can be sure that the Apicius recipe represents the first French toast. But it is clear that similar bread dishes were popular throughout Medieval Europe, albeit with varying names. In Germany, the recipe was called *arme ritter;* in England, *suppe dorate;* and in Portugal, *fatias douradas,* which literally translates to "golden slices of bread."

Interestingly enough, if you go to France, you won't find French toast, or even *Toast à la Française,* on the menu. Their version of eggy bread is called *pain perdu* ("lost bread"), a likely reference to the reclamation of stale bread that would otherwise be lost to the garbage or the pigeons. In the United States, French-speaking cooks in Cajun areas of Louisiana make *pain perdu* using thick slices of local French bread and a topping of cinnamon, powdered sugar, and Louisiana cane syrup.

The French might lay claim to bringing the French toast recipe to the United States, but some people offer up an entirely different explanation for the name. According to this alternate theory, French toast is an all-American creation, first made by Joseph French at his roadside tavern in Albany, New York, circa 1724. Though Mr. French wished to credit himself for the sweet, golden breakfast treat he supposedly invented, he listed it on the menu as French toast—not French's toast.

Why? Because he didn't know how to use an apostrophe.

So does French toast have its roots in ancient culinary history or an unfortunate grammatical oversight? When you have a plate that's piled high with pan-fried toast, powdered sugar, strawberries, and syrup, who really cares?

Q Why is Maine called "Down East"?

A The state of Maine occupies the northeastern-most corner of the United States. You might think, therefore, that when people in Boston, which is fifty miles to the south, take a jaunt to Maine, they would say they're going "up to Maine" for the summer and returning "down to Boston" when the season is over. Instead, Bostonians, or at least the old-fashioned "proper" ones, will tell you that they're going "down to Maine" and coming back "up to Boston." Say what? Have they lost their compass?

Not exactly. The phrase "down east" comes from sailors' lingo. Back in the nineteenth century, the fastest way to travel was by clipper ship. Fortunately, a steady wind from the south swept up the East Coast, pushing ships northeast. When sailors travel with the wind at their backs, they say they are traveling downwind. "Down east" means going east with the wind behind you. Returning south, ships would be pushing against the wind, or upwind.

Pretty simple. But in this day and age, when most vacationers arrive by interstate highway, why do Mainers still like to call their state Down East? Maybe it's because people who can stick it out in a land of long, dark winters are pretty darn proud of their history and like to celebrate it in all kinds of unique ways, from choosing the white pine cone as their state "flower" to claiming Moxie as their official state drink.

And, of course, nothing's more fun than confusing first-time tourists with friendly signs pointing them north to Down East.

Q Why is the same product called different things in different countries?

A Some product names travel well—they're the same whether you're speaking English, Spanish, or some other language. In other cases, a product name that's perfect for one market runs into...issues...in other markets. The meanings of many foreign words get lost in translation when converted into English, and vice versa. For example, the English names of the following products would probably benefit from a name change if they want to be successful in the United States.

Cream Collon: Glico's Cream Collon is a tasty cookie from Japan. The small cylindrical wafers wrapped around a creamy center actually do resemble a cross section of a lower intestine filled with cream. An ad says to "Hold them between your lips, suck gently, and out pops the filling." Yum! Glico's Cream Collon can be ordered on the Internet.

Ass Glue: Ass glue is made from fried donkey skin and is considered a powerful tonic by Chinese herbalists, who use it to fortify the body after illness, injury, or surgery. If you have a dry cough, a dry mouth, or are irritable, you can find ass glue at most Chinese herb shops.

Mini-Dickmann's: Mini-Dickmann's, a German candy made by Storck, is described as a "chocolate foam kiss." Available in milk, plain, or white chocolate, Mini-Dickmann's are only an inch and a half long. Too embarrassed to be seen with a box of Mini-Dickmann's? Try Super-Dickmann's, the four-inch variety. Both sizes are available from Storck USA.

Kockens Anis: If you think anis sounds funny, you'll laugh even harder when you see Kockens Anis in Swedish grocery stores. Anis is aniseed, a fragrant spice used in baking, and Kockens is the brand name. While aniseed is found in most U.S. grocery stores, don't ask for the Kockens brand because it's not available and could get you coldcocked.

Aass Fatøl: On those rare hot days, Norwegians like to quench their thirst with a cold bottle of Aass Fatøl beer. The word fatøl appears on many Scandinavian beer labels and means "cask." This beer comes from the Aass Brewery, the oldest brewery in Norway. If you'd like to get some Aass, it's imported in the United States.

Big Nuts: Big Nuts is a chocolate-covered hazelnut candy from the Meurisse candy company in Belgium. For those who like candy that makes a statement, Big Nuts is available online.

Dickmilch: Dairy cases in Germany are the place to find dickmilch, a traditional beverage made by Schwalbchen. In German, dickmilch means "thick milk" and is made by keeping milk at room temperature until it thickens and sours. Called sour milk in the United States, it's a common ingredient in German and Amish baked goods.

Pee Cola: If you're asked to take a cola taste test in Ghana, one of the selections may be a local brand named Pee Cola. The drink was named after the country's biggest movie star Jagger Pee. Don't bother looking for a six-pack of Pee to chug because it's not available in the United States.

Piddle in the Hole: Take a Piddle in the Hole at a pub in England and you'll be drinking a beer from the Wyre Piddle Brewery. The same brewery in the village of Wyre Piddle also makes Piddle in the Wind, Piddle in the Dark, and Piddle in the Snow. Sadly, Piddle is only available in the UK.

Shito: Shito is a spicy hot chili pepper condiment that, like ketchup in the United States and salsa in Mexico, is served with most everything in Ghana. There are two versions: a spicy oil made with dried chili pepper and dried shrimp; and a fresh version made from fresh chili pepper, onion, and tomato. Shito appears as an ingredient in Ghanaian recipes but hasn't found a market in the United States.

Fart Juice: While it may sound like an affliction caused by drinking it, Fart Juice is a potent potable in Poland. Made from the leftover liquid from cooking dried beans, this green beverage could pass for a vegetable juice and is probably a gas to drink, but it's not available in the United States.

Q Why isn't Scotland Yard in Scotland?

A British nomenclature is loaded with misleading terms. Plum pudding is not pudding, nor does it contain plums. Real tennis doesn't have much to do with real tennis. Spotted dick is not a venereal disease, but rather a delicious dessert—a dish that thankfully has nothing to do with what you might think.

Given this legacy of verbal imprecision, it's perhaps not surprising that the headquarters of the famous police force that patrols London is called Scotland Yard.

It started in 1829, when Charles Rowan and Richard Mayne were charged with organizing a citywide crime-fighting force. At the time the two men lived together in a house at 4 Whitehall Place, and they ran their fledgling outfit out of their garage, using the back courtyard as a makeshift police station. "Rowan and Mayne's Backyard" wasn't an appropriate name for the headquarters of a police force. Instead, it was called Scotland Yard. Why? London police don't play bagpipes; haggis isn't part of the rations; and as far as we know, London policemen don't wear kilts (at least not in public). So what gives?

You might think that a mystery like this would be a perfect case for Scotland Yard. Unfortunately, those famed investigators aren't sure how their hallowed institution got its name, which is not necessarily a compelling endorsement of their detective work. After years of research, though, word detectives have narrowed the origin of the name to two likely possibilities.

According to the first explanation, Scotland Yard sits on the location of what was once the property of Scottish royalty. The story goes that back before Scotland and England unified in 1707, the present-day Scotland Yard was a residence used by Scottish kings and ambassadors when they visited London on diplomatic sojourns.

The other, less regal possibility is that 4 Whitehall Place backed onto a courtyard called Great Scotland Yard, named for the medieval landowner—Scott—who owned the property.

Regardless of the name's true origin, the Metropolitan police have moved on—sort of. In 1890 they decided that they needed new digs and moved to a larger building on the Victoria Embankment. Given a chance to redeem themselves and give their headquarters a name that actually made sense, what did the London police choose?

You guessed it. New Scotland Yard.

Q Could New York have been Nouvelle-Angoulême?

A In 1524, the famed Italian explorer Giovanni da Verrazzano brought Europe its first eyewitness description of New York Harbor and its friendly natives. His new

York stopover was part of a lengthy seagoing effort to find the ever-elusive passage to the Pacific Ocean for King François I of France. What land should France claim? Only by knowing the sea passages could François decide.

The expedition originally included four ships and departed Normandy in 1523. Two of the original ships turned back off Brittany with storm damage, however, and one went privateering. Only the multimast carrack *La Dauphine* made it to the Portuguese archipelago of Madeira, where it wintered and took on supplies for the next leg of the voyage across the Atlantic.

Aboard *La Dauphine,* the expedition made its first North American landfall in early spring 1524 off modern-day North Carolina and then followed the coast northeast. Da Verrazzano somehow missed Chesapeake Bay on his way to modern New York Harbor.

The explorer's entry into the harbor indeed passed through the narrows (yes, those with the bridge bearing his name today). Loath to risk *La Dauphine* in tricky exploration of the shoreline, however, da Verrazzano set out across Upper New York Bay in a small boat. He mistook the Hudson for part of a lake and didn't explore further, but he did meet with cordial Lenape Native Americans. His recollections about their dress and agriculture remain valuable even today.

Da Verrazzano continued northeast to what later became New England and finally returned to France in July. He named the NYC area *Nouvelle-Angoulême,* which means New Angoulême. (Didn't stick, but nice try.) "Old" Angoulême, by the way, is a pleasant but unremarkable region in southwestern France about an hour's drive east of Bordeaux.

Da Verrazzano's quick-and-dirty tour of the modern northeast U.S. coast eventually faded into obscurity, which is unfortunate given its pioneering nature. The explorer glossed over numerous inland waterways (none of which actually led to China, of course, but he failed to confirm this). Unluckiest of all, da Verrazzano had the misfortune to operate shortly after Cortez and Magellan, whose feats upstaged his.

Had he been thorough and written more, we might remember da Verrazzano differently. But as New York's first European visitor, a distant kinsman of some who would one day call the city home, Giovanni da Verrazzano has a secure parking place in New York's history.

Here's some other tidbits about his trip and its aftermath:

- The Abenaki American Indians of Maine were far less friendly than those around New York Harbor. They mooned da Verrazzano.

- His brother Girolamo, brought along as cartographer, jumped to the conclusion that North America was two halves divided by (what we now call) a mythical "Sea of Verrazzano." The error wasn't cleared up for more than a century.

- Two other bridges are named for da Verrazzano: one in Rhode Island's Narragansett Bay and one connecting Assateague Island to the mainland of Maryland.

- The correct spelling of the explorer's name is definitely "da Verrazzano." The I-278 bridge that stands in his honor spells the name wrong, as does the bridge in Maryland. Rhode Island's bridge spells it correctly, though everyone leaves out the "da," which is like referring to the famous NYC mayor as "Fiorello Guardia."

• Some believe da Verrazzano's letter to King François I describing the voyage was a fake and that he never came to America at all. However, most historians accept his travels as genuine.

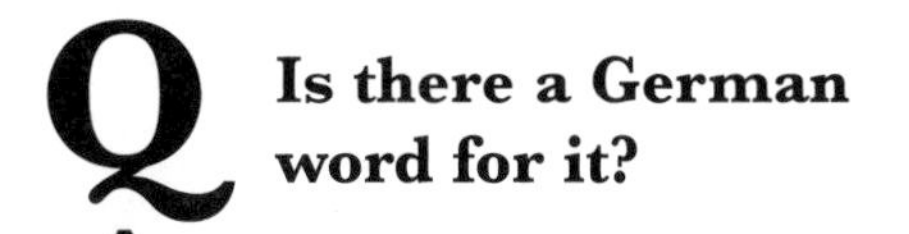

Is there a German word for it?

A Many everyday words in the English language came from German. In fact, they're so familiar that we rarely think of their German roots. Some such words include *blitz, bratwurst, delicatessen, diesel, kindergarten, pretzel, poltergeist, pumpernickel, rucksack,* and *zeitgeist,* among others.

You may also recognize *schadenfreude,* which is the pleasure one feels when something bad happens to someone we dislike. (There's a related word, *freudenschade,* which was modeled after the German word but doesn't exist in German. It names the feeling of being miserable because of someone else's success.)

But there are plenty of funny-sounding words borrowed into English from German. These may bring pause to a conversation because they're not exactly words you hear every day: *Torschlusspanik* is a such a word, for example. It means "a sense of anxiety, especially in middle age, caused by feeling that one no longer has the same opportunities as one did when younger." It comes from German words that literally mean "gate-shut panic," as if you were seeing a door close on you.

There are also such wonderful German words as *zugzwang,* a situation in a chess game where a player must make an undesirable move; *zugunruhe,* the urge to migrate, especially in captive birds; and *treppenwitz* (what the French call *l'esprit d'escalier,* or "staircase wit"), which is the witty, funny, or cutting thing you think of to say after you have already left the conversation (or the party).

A couple lesser-known German words that have been adopted by the occasional English speaker are *schlimmbesserung,* which is the act of making something worse while trying to make it better, and *fingerspitzengefühl,* which names a good sense or intuition about how things (usually delicate or difficult things) ought to be done.

After reading this article and learning new words, it might not be a stretch to say you have or are beginning to develop *sprachgefühl,* which literally means "a feeling for language" and is used when someone has a good ear for language and knows what sounds correct or appropriate.

Q What are some other perfect words from other languages?

A English borrows freely from nearly every language it comes into contact with. There are still, however, many concepts and situations for which Anglophones lack *le mot juste* (the perfect, precise word, from the French). Here are some suggested foreign words to add to the English dictionary, because you never know when you'll need them.

backpfeifengesicht (German): a face that's just begging for someone to slap it (a familiar concept to anyone fond of daytime television)

bakku-shan (Japanese): a girl who looks pretty from the back but not the front; this loanword would in fact be a loanword regifted, since it's already a combination of the English word back with the German word *schoen,* meaning "beautiful"

blechlawine (German): literally "sheet metal avalanche": the endless line of cars stuck in a traffic jam on the highway

bol (Mayan): for the Mayans of southern Mexico and Honduras, the word bol pulls double duty, meaning both "in-laws" and "stupidity"

drachenfutter (German): literally "dragon fodder": a makeup gift bought in advance; traditionally used to denote offerings made by a man to his wife when he knows he's guilty of something

dugnad (Norwegian): neighborly mutual aid; for instance, used to describe a party where the object is to help the host with something work-intensive, like painting the house or moving

ilunga (Tshiluba, Democratic Republic of the Congo): a person who is ready to forgive any abuse for the first time, to tolerate it a second time, but never a third time

karelu (Tulu, south of India): the mark left on the skin by wearing anything tight

kummerspeck (German): literally "grief bacon": excess weight gained from overeating during emotionally trying times

Ølfrygt (Viking Danish): the fear of a lack of beer; often sets in during trips away from one's hometown, with its familiar watering holes

putzfimmel (German): a mania for cleaning; indispensable for people with neat-freak significant others or roommates

uitwaaien(Dutch): walking in windy weather for the sheer fun of it

Q Why is coffee called "joe"?

A With the exploding popularity of gourmet coffee drinks in recent years and the vast number of specialty, fair-trade, and organic coffee purveyors now dominating the market, it's sometimes a challenge to find a joint that serves up plain old coffee. And when you do stumble upon such a place, asking for a "cuppa joe" is more likely to be met with a blank stare than a cup of coffee (unless the barista is a fan of old, hard-boiled detective movies). Yet for much of the twentieth century, coffee was indeed referred to as joe.

Why "joe"? Why not Bob? Fred? Orville? There are a number of prevailing theories as to why coffee is referred to as joe. The first, and the one promoted by the United States Navy, holds that in 1913, new Navy Secretary Josephus Daniels abolished the policy of allowing sailors to drink alcohol at mess. In false praise, American sailors began referring to coffee—now the most powerful beverage available to them—as a "cup of joe." However, most etymologists discard this theory, pointing out that the first time that the phrase "cup of joe" appeared in print was in 1930, and a seventeen-year gap between the first colloquial use and the first recorded use is virtually unheard of.

A second explanation is only partly military in origin. The term "joe" long referred both to an average American and to an American soldier (think G.I. Joe), and because coffee is both the average man's drink of choice and a primary staple of a soldier's rations, coffee became associated with the name Joe. This makes sense, right?

The least interesting theory (but the one that's most likely correct, according to some

etymologists) suggests that "joe" is a bastardization of Java, the island that for a long time was the primary source of coffee to North America.

Of course, those folk pronouncing Java as "joe" might have been drinking something a lot stronger than joe. Jack, perhaps.

FOOD AND DRINK

Q Did the Italians steal their spaghetti?

A Talk about a raw deal. Almost as soon as we learn to love spaghetti and any number of other pasta staples, we're told to hold up—the Italians didn't invent spaghetti. Innocence can end in a variety of ways, and for some of us, it's when we're told that pasta pitchers like Al Molinaro, of *Happy Days* and On-Cor commercial fame, have been lobbing us a lie.

The "untold story" usually centers on Marco Polo, the Italian explorer who filled a bag full of Eastern innovations and then sailed back to Venice in the late thirteenth century. While it's true that the Chinese made noodles for thousands of years before Italians had the bright idea of dribbling tomato sauce on them, Chinese noodles are made from rice, whereas spaghetti comes from wheat.

But the Italians aren't off the hook, because while they didn't steal pasta from the Chinese, they likely did pilfer it from their then-enemies to the south, the Arabs. Pastas that we consider Italian—such as spaghetti—were actually brought to the country by Arab conquerors in the mid-twelfth century. In fact, spaghetti can be traced through Aramaic writings all the way back to the fifth century; it's even mentioned in the Jerusalem Talmud, which indicates that spaghetti was present in ancient Jerusalem.

Arab pasta makers turned standard dumplings (imagine the gnocchi of today) into long, thin strands of spaghetti. Because it could be dried and stored for months or even years, spaghetti was a food innovation rivaling rice for flexibility and shelf life.

Never ones to sleep on a good idea, the Italians can at least be credited for the mass production of pasta. Italians built the first pasta-making factories in 1824, which makes them directly responsible for turning Wednesday into Prince Spaghetti Night.

Q Are Rocky Mountain oysters an aphrodisiac?

A For thousands of years, people from every culture have sought to inspire or enhance the act of making love. Herbs, potions, and animal parts that look like human genitalia are among the substances that people have consumed, snorted, and rubbed on themselves in the quest for an effective aphrodisiac.

Among the most obvious and primitive aphrodisiacs are the reproductive organs of animals that are considered to be especially virile. In the American West, the-hope-to-be-hot assign erotic properties to bull testicles, delicately euphemized as "Rocky Mountain oysters."

Testicles are cut from young bulls to render the ornery beasts more docile and, thus, easier to raise for food. The soft, slimy meat is considered a delicacy. Sliced, breaded, and fried, it's served up as Rocky Mountain oysters or, if you prefer, Montana tendergroins, cowboy caviar, or swinging beef.

Brave and hungry souls—some undoubtedly possessed of the urge to merge—can be found chowing down at "testicle festivals." Many say that Rocky Mountain oysters taste like chicken; others liken the flavor to fried shrimp or marine oysters; and some maintain that the only part they can taste is the breading.

As for their power to prime the love pump? The U.S. Food and Drug Administration turns a cold shower on the notion, declaring that there's no scientific proof of the effectiveness of any reputed aphrodisiac, bull-based or otherwise. But if you think that eating a longhorn's privates can help, go ahead and have a ball.

Q What is head cheese?

A The name of this delicacy is deceptive, because head cheese has absolutely nothing in common with your favorite mozzarella, Cheddar, or Brie. Head cheese isn't a dairy product at all—it's a jellied loaf of sausage. If you want to get fancy, you can even call it a terrine.

As for the head part, that's right on. Head cheese is made with meaty bits from the head of a calf or pig, or sometimes even a sheep or cow. That's the traditional recipe, anyway; today's head cheese might include other edible animal parts, including feet, tongues, and hearts.

Getting back to that head, it's usually split or quartered and simmered in a large stockpot until the meat becomes tender and falls off the bone. Any meat remaining on the skull is picked off, and then it's all chopped up.

At this point, seasonings are added. Ingredients vary by culture, region, or even butcher. In Denmark, head cheese *(sylte)* is spiced with thyme, allspice, and bay leaves. In southern Louisiana, where it's also known as "souse," head cheese is traditionally flavored with vinegar and hot sauce.

What really makes head cheese come together is the cooking liquid in the stockpot. As the calf or pig or sheep head simmers, the collagen from the skull cartilage and marrow leeches into the broth. This collagen-infused stock is added to the chopped, seasoned meat, and the whole mixture is poured into a pan or mold. From here, the head cheese is cooled in the refrigerator, and voilà—the collagen causes the mixture to set and solidify into a gelatin.

At this point, the head cheese is ready to be removed from the mold. Usually served at room temperature, it can be thinly sliced and eaten with crackers, or cubed like cheese for a tasty appetizer. Look for it ready-made at your neighborhood deli or supermarket—and be sure to serve it to your most deserving guests.

Q What is the oldest still-popular alcoholic beverage?

A That would be beer, the alcoholic beverage of choice for millions of thirsty folks dotting the globe.

Beer dates back at least to the sixth millennium B.C. in Mesopotamia, a region located for the most part in what is today Iraq, which brings up a logical question: Why do humans still drink zestless Bud Light in mass quantities despite the fact that they have had more than seven millennia to refine their taste in beer? Perhaps the Sumerian goddess Ninkasi would know: The Sumerian hymn to her is also a recipe for beer itself.

It's been speculated that beer was discovered by accident, when some bread dough was left out in a Sumerian courtyard, was rained on, and fermented into a beerlike liquid over the next several days. If that's true, it's a charming bit of good luck for humanity. But even more thought-provoking is this golden nugget of information: It is widely believed that bread and beer were the catalysts for civilization.

Until humans discovered that certain grains could be made into bread and beer, they were nomadic wanderers. Once they lucked into the earliest ancestors of Wonder Bread and Pabst Blue Ribbon, they settled down into communities that shared efforts in cultivating and selling grain, and produced food on the spot. In this epochal light, the annual Super Bowl "Bud Bowl" ads seem a bit shallow, no?

Anyhow, despite its exalted position in human history and football, beer isn't the oldest known booze. That would be a Chinese rice wine sweetened in the making with fruit and honey. It was discovered in 2004, its molecules clinging to nine-thousand-year-old pot fragments from a Stone Age site in northern China. In other words, it's probably about a thousand years older than beer.

Take that, Augie Busch.

What's so hot about a hot dog?

A Honestly? Not much. The "hot" in hot dog doesn't mean spicy, sexy, stolen, or anything that super exciting. It's just a reference to the way the sausages are served. And that would be warm, and preferably in a soft, sliced bun.

What really gets people hot—as in, riled up—is how the term "hot dog" came about. It's agreed that the hot dog, a.k.a. frankfurter, originated in Germany. Some say it was first made in Frankfurt am Main around 1484; others insist it was in Coburg in the late sixteen hundreds. At any rate, what really matters is that the Germans referred to their skinny sausage creation as "dachshund" or "little dog" sausage. This was most certainly a nod to their country's popular short-legged, long-bodied dog breed.

Dachshunds also go by the nickname of "wiener dogs," and whaddaya know—hot dogs are known as wieners, too. And no, this doesn't mean that hot dogs are made from dog meat. American hot dogs can be crafted from beef, pork, veal, chicken, turkey, or any or all of the above.

Now we know that "dog" was a longtime common synonym for sausage, but just when did the "hot" come into play? Many sources credit American journalist and cartoonist Thomas Aloysius Dorgan, or TAD, for coining the term. (This guy is also credited with coming up with such phrases as "the cat's meow" and "for crying out loud.") On a chilly day at New York's Polo Grounds in April 1901, according to the National Hot Dog & Sausage Council, vendors weren't making any money on the usual frozen ice cream and cold soda, so they started selling dachshund sausages from portable hot water tanks. The sales pitch went something like this: "They're red-hot! Get your dachshund sausages while they're red-hot!"

As a sports cartoonist for the *New York Journal,* Dorgan took in the spectacle and conjured a caricature of barking dachshund sausages that were warm and comfortable in sandwich rolls. However, he didn't know how to spell "dachshund," so he wrote "hot dog!" instead. The cartoon was apparently so popular that the term "hot dog" entered the culinary lexicon.

The problem is, historians have never been able to dig up a copy of Dorgan's supposed "hot dog!" cartoon. And this has a lot of people shouting, "Bologna!" In fact, many experts, including recognized hot dog historian Bruce Kraig, say that the term "hot dog" was appearing in college magazines by the 1890s.

So maybe what makes a hot dog hot is not so much its temperature, but rather the amount of heated debate that surrounds it.

Q Is black pudding good for dessert?

A Maybe if you're Dracula or a parasitic leech. Black pudding isn't a sweet, creamy, chocolaty custard made by Swiss Miss. In England, Ireland, and Scotland, it's part of a traditional full breakfast. So no, you probably wouldn't want to trade your slice of apple pie for a large-link sausage that is made with pig's blood.

In North America and other parts of the world, black pudding is known by a more conspicuous name: blood sausage. To make it, fresh pig's blood is combined with suet (that's the hard, white fat from the pig's kidneys and loins), breadcrumbs, oatmeal, and seasonings (usually black pepper, cayenne, mace, coriander, herbs, and onion). This mixture is cooked together, stuffed into sausage casings (that's a nice way of saying "intestines"), and lightly poached.

From whose twisted mind did this sanguinary side dish sprout? Many food historians think black pudding has its origins in ancient Greece. Homer's Odyssey, written around 800 B.C., even makes poetic reference: "As when a man beside a great fire has filled a sausage with fat and blood and turns it this way and that and is very eager to get it quickly roasted . . ."

The oldest documented recipe for black pudding is attributed to Apicius, a collection of Roman cookery recipes from the first few centuries A.D. In this version, the blood is mixed with chopped hard-boiled egg yolks, pine kernels, onions, and leeks.

It's likely that most black pudding recipes came from the economic need to make use of everything when a pig was butchered. In medieval Europe, even relatively poor families had a pig for the annual late-autumn slaughter. This is a possible reason why black pudding became a delicacy to be enjoyed on the feast days.

Today, black pudding can be purchased already prepared (no need to slaughter a pig in your own backyard) and can be enjoyed any time of year. It only requires a gentle reheating in the fry pan, grill, or oven. Serve it sliced alongside fried eggs and bacon for a traditional UK breakfast, or try a new-fangled gourmet rendition along the lines of black pudding with wild mushroom sauce.

However elegant the preparation, the thought of black pudding may still make you squeamish. In this case, you might opt for a white pudding instead. Similar to black pudding (but sans the plasma), this sausage is made with white meat (chicken or pork), fat, oatmeal, and seasonings. Just steer clear of the original recipe—very old versions of Scottish white pudding call for sheep brain as a binding agent.

Q Is there a killer sushi?

A If you're planning to have a dignitary from Japan over for dinner, there's one delicacy from his homeland you may want to avoid preparing: pufferfish. The pufferfish, also known as blowfish or fugu, is a homely creature that, when threatened, inflates itself and displays protective spikes that are filled with tetrodotoxin, a neurotoxin that is about 1,200 times more deadly than cyanide. The average pufferfish has enough of it in its three-foot-long body to kill thirty people.

Believe it or not, pufferfish is served raw as sushi, after the tetrodotoxin has been removed. This is, however, an inexact science; about one hundred people die every year in Japan from pufferfish that have been improperly prepared. The initial symptom of pufferfish poisoning is paralysis of the lips and face, which can appear from ten minutes to several hours after ingestion. The cause of death is respiratory paralysis. There is no known antidote to tetrodotoxin, but the treatment of symptoms includes aggressive measures to keep the airways open.

Sushi chefs who want to work with pufferfish go through an intensive program of study in Tokyo. They're taught how to prepare the creature for consumption, including how to cut and separate the toxic parts from the edible ones. Last, but certainly not least, they're taught first aid.

Why would someone eat pufferfish? Well, it's akin to mountain climbing, bungee jumping, or skydiving—the thrill of trying to cheat death. When a person at a sushi bar orders pufferfish, it is traditional to offer many toasts to his or her health. This person, this gastronomic renegade, becomes the center of attention.

While the pufferfish is an extreme example, sushi in general is a relatively high-risk food. Raw fish is full of bacteria, and mercury levels—particularly in tuna—have become an issue. The traditional accompaniments to sushi are meant to help, not just add flavor. Vinegar is added to the rice to heighten the pH level and potentially kill bacteria; wasabi and pink pickled ginger also have bacteria-killing properties.

Nevertheless, you might want to consider introducing that Japanese dignitary to a dish called pizza.

Q Couldn't the Irish have found something to eat besides potatoes to avoid a famine?

A More than a million people starved to death in Ireland from 1845 to 1851. The disaster is called the Great Famine, but it wasn't really a famine. Only one crop failed: the potato. How could this have killed so many? Why didn't the Irish eat cabbage or scones or even chalupas, for crying out loud?

The answer is simple: Those who starved were poor. For generations, the impoverished in Ireland had survived by planting potatoes to feed their families. They had nothing else. Ireland's wealthy landowners grew a wide variety of crops, but these were shipped away and sold for profit. Most of the rich folks didn't care that the poor starved.

How did things get so bad? Irish History 101: The Catholics and the Protestants didn't like each other, and neither did the English and the Irish. Back then, the wealthy landowners were mostly Protestants from England, while the poor were Catholic peasants. The Irish peasants grew their food on small parcels of land that were rented from the hated English.

In the sixteenth century, a hitherto unknown item crossed the Atlantic from Peru, originally arriving in England and finally getting to Ireland in 1590: the potato. Spuds grew well in Ireland, even on the rocky, uneven plots that were often rented by peasants, and they quickly became the peasants' main food source. Potatoes required little labor to grow, and an acre could yield twelve tons of them—enough to feed a family of six for the entire year, with leftovers for the animals.

We think of potatoes as a fattening food, but they're also loaded with vitamins, carbohydrates, and even some protein. Add a little fish and buttermilk to the diet, and a family could live quite happily on potatoes. Potatoes for breakfast, lunch, and dinner might sound monotonous, but it fueled a population boom in Ireland. By the nineteenth century, three million people were living on the potato diet.

In 1845, though, the fungus *Phytophthora infestans,* or "late blight," turned Ireland's potatoes into black, smelly, inedible lumps. Impoverished families had no options, no Plan B. Their pitiful savings were wiped out, and they fled to the work houses—the only places where they could get food and shelter in return for their labor.

When the potato crop failed again the next year, and every year through 1849, people began dying in earnest—not just from starvation, but from scurvy and gangrene (caused by a lack of vitamin C), typhus, dysentery, typhoid fever, and heart failure. Overwhelmed and underfunded, the work houses closed their doors. Many people who were weakened by hunger died of exposure after being evicted from their homes. To top the disaster off, a cholera epidemic spread during the last year of the blight, killing thousands more.

The exact number of those who perished is unknown, but it's believed to be between one and two million. In addition, at least a million people left the country, and many of these wayward souls died at sea. All during that terrible time, plenty of food existed in Ireland, but it was consumed by the wealthy. The poor, meanwhile, had nothing. They were left to starve.

Q What's the difference between brandy and cognac?

A Cognac is to brandy what champagne is to sparkling wine. Does that help? If not, try this: More than anything, the distinction between cognac and brandy is geographical.

Cognac is a type of brandy that is made exclusively from the grapes that grow in a specific region of France. Connoisseurs say that cognac is perhaps the finest of all brandies. The clerk at the corner liquor store, meanwhile, is more concerned about the fact that it's the most expensive brandy you can buy; it's behind the counter, so please ask nicely.

Brandy is no more nor less than distilled, fermented fruit juice. Anything that's simply called "brandy" is made from fermented grapes, like wine. When brandy is made from other fruits, it's indicated in the name. An example is apple brandy, which is produced from cider.

As one of the earliest forms of distilled wine, brandy has a distinguished place in the history of spirits. Distilled wine was the original hard liquor, and it was popularized by the court physicians of Renaissance-era Europe (who thought it had medicinal properties). They got the idea of distillation—which purifies thedrink and increases its alcohol content—from Arab alchemists.

The word "brandy" itself derives from the Dutch *brandewijn* ("burnt wine"). It has been widely enjoyed for more than five hundred years, and (as the story goes) was what was carried around by Saint Bernard dogs in tiny kegs in the Swiss Alps. But you don't need to be snowbound to enjoy its warming properties. So in the words of the poet Busta Rhymes, "Pass the Courvoisier."

Q What's the unhealthiest dish ever concocted?

A You probably wanted us to conduct serious research into this one—maybe some double-blind studies, perhaps a bunch of empirical data. That would be neat. But, to paraphrase *Animal House's* Otter when he's contemplating an assault on the entire Faber College Greek system, it would take years and cost millions of lives.

Besides, a stupendous effort is unnecessary when there's the Hamdog.

The Hamdog was created by a bar owner in Decatur, Georgia. It starts as a hot dog wrapped in a hamburger patty. It's deep-fried, smothered in chili, cheese, and onions, and served on a hoagie bun. Oh, by the way, it's topped with a fried egg and a pile of French fries. The same guy invented the Luther Burger, a bacon cheeseburger served on a bun fashioned from a Krispy Kreme doughnut. Luckily for humanity, the bar has since closed.

A report on the Hamdog said the burger and hot dog alone comprise eighty-five grams of fat—well above the average person's recommended daily intake of sixty-five grams. Factor in its other ingredients and consider the fat that's absorbed in the frying process, and the Hamdog might deliver a week's worth of dietary fat—much of it the bad kind. And that's not even considering its cured-meat chemicals and other bad molecules.

The bottom line? The Hamdog packs enough artery-hardening punch to earn the "unhealthiest" prize in our book, particularly since we're suddenly too queasy to think of an alternative.

Q Why do doughnuts have holes?

A The saga of how doughnuts came to have holes is a bit of a mystery; perhaps a police detective is needed to solve it. What classic cop wouldn't want to pore over mountains of evidence that involves doughnuts?

The origin of doughnuts most likely can be traced to Northern Europe during medieval times. Called *olykoeks* ("oily cakes"), the pastries came to America with the Pilgrims, who had picked up the recipe in Holland, their first refuge from England, which they abandoned for America in the early sixteen hundreds. The dough in the middle of these pastries rarely got cooked, so that area often was filled with apples, prunes, or raisins.

By the mid-eighteen hundreds, the pastries were being made with a hole in the middle—and this is where the plot thickens. Two stories about the origin of the hole involve Hanson Crockett Gregory, a sea captain from Rockport, Maine. One says that he poked out the middle of one of his wife's homemade doughnuts by plunging it into a spoke on the ship's wheel. That eliminated the uncooked middle, and it enabled Gregory to eat and keep his boat at an even keel at the same time.

A second story—this one slightly more plausible—involves Gregory eating doughnuts with other crew members. Tired of the raw dough in the middle, he took a tin off the ship's pepper box and used it to push out the middle, leaving only the cooked edges. He tasted it and exclaimed that it was the best doughnut he had ever eaten. Years later, in 1916, Gregory recounted this story in the *Washington Post.*

There is no real proof that backs up either account involving Gregory, but this much is certain: A plaque commemorating his culinary claim stands at the house in Maine where he lived. And perhaps not coincidentally, doughnuts did indeed have holes by the mid-eighteen hundreds, making them easier to cook and improving their taste.

Once they started coming with holes in them, doughnuts soared in popularity. During WWI, the French gave doughnuts to American soldiers to remind them of home. In the 1920s, doughnuts were the snack of choice in movie theaters. At the 1934 World's Fair in Chicago, they were called, "The food hit of the Century of Progress." Cops all over America couldn't agree more.

Q Why did we start mixing other stuff with hard liquor?

A Yankee ingenuity? Or maybe it was our sweet tooth. While straight booze was historically the beverage of the masses, mixed drinks with lots of sugar added were reserved for wealthy landowners.

In 1806, a Hudson, New York, newspaper defined a cocktail as stimulating liquor combined with sugar, spirits of any sort, bitters, and water. This was one of the first recorded uses of the word "cocktail" in print. No one can be sure where the word originated, though there are many anecdotes. Colonial Americans drank beer, wine, cider, and rum, and sometimes mixed rum into punch. By the late seventeen hundreds, visitors to the United States observed that Americans liked to start the day with a "sling": a drink of strong spirits, sugar, and bitters or herbs. The mint julep was a sling. Men often downed several slings before lunchtime.

Cocktails were mentioned by early American writers, including Washington Irving and James Fenimore Cooper, author of *Last of the Mohicans*. A cocktail recipe book appeared in the 1860s. The first martini recipe was printed in 1884, though the ingredients were not what we'd put into a martini today—early martinis were sweet, not dry.

Sweet alcoholic drinks were mixed in the home, at parties, and sometimes in fine hotels. Saloons did not serve them: They sold only straight liquor and beer, not cocktails, until sometime in the 1880s. Mixed drinks then became so popular that saloon owners became mixologists, and cocktails went on to enjoy a golden age of innovation.

Prohibition ended this golden age in 1920. Selling alcohol was suddenly illegal, so the trade moved underground. This hardly stopped consumption, though. Adding soft drinks or milk stretched the precious bootlegged whiskey and masked the vile taste of bathtub gin. Cocktails remained in high demand and were sold in speakeasies during Prohibition; ever since, they've been hawked at bars, clubs, and restaurants.

Journalist H. L. Mencken contended that the cocktail is "the greatest of all the contributions of the American way of life to the salvation of humanity." No telling how many "contributions" he had enjoyed on the night he wrote that.

Q How sweet are sweetbreads?

A A tip for those who rarely eat at trendy restaurants: If you see sweetbreads on the menu, don't start salivating at the thought of a warm muffin with butter dripping down the sides. Instead, picture the thymus gland or pancreas of a young sheep, cow, or pig. Then exhale deeply and start focusing on taking a swig or two from your glass of wine.

Sweetbreads are a delicacy enjoyed throughout the world by people with adventurous palettes, but the burger-and-fries types might not understand such culinary wanderlust. In fact, they might want to ask the question: What in the name of the Golden Arches is a thymus gland? The answer isn't pretty. A thymus gland contains two lobes, one in the throat and the other near the heart. The lobe near the heart—particularly from milk-fed young calves—is considered the best to eat because of its smooth texture and mild taste; as a result, it will cost you more at that trendy restaurant. Pancreas sweetbreads, or stomach sweetbreads, are much less common than their thymus counterparts.

Sweetbreads and other edible internal organs are often grouped together using the term "offal" (which, for those still ready to vomit, isn't a word for "awful" in some foreign language). It means the "off-fall," or the off-cuts, of a carcass.

Since sweetbreads aren't sweet and aren't bread, how did they get their name? This is something of a mystery. *The historie of man,* published in 1578, sheds a splash of light on the matter: "A certaine Glandulous part, called Thimus, which in Calues . . .is most pleasaunt to be eaten. I suppose we call it the sweete bread." Translation: They tasted good.

Back in those roughhewn days—before butcher shops and grocery stores—sweetbreads weren't considered a delicacy. Families butchered their own livestock and often ate every part, including the thymus gland and pancreas. Today, sweetbreads are prepared in many ways: You can poach, roast, sear, braise, or sauté 'em, and often season them with salt, pepper, onions, garlic, or thyme.

If you want to prepare sweetbreads, we have two pieces of advice. First, sweetbreads are extremely perishable, so be sure to cook them within twenty-four hours of your purchase. Second, they're probably not the ideal dish to serve on a first date.

Q Is there any duck in duck sauce?

A If you want to keep things authentic, the question should be, "Are there any plums in plum sauce?" because duck sauce is a nickname that plum sauce adopted over time. However, when speaking of Chinese food, there seems to be very little authenticity left.

Yong Chen, a history professor at the University of California-Irvine, says that Chinese food "is quintessentially American." Ever since the first Chinese restaurants opened in California mining towns in the mid-nineteenth century, Chinese restaurateurs have looked for ways to Americanize their dishes. In China, and in American fine-dining establishments that serve Peking duck, the traditional sauce that accompanies this dish is hoisin sauce (which is soy-based), not plum sauce.

However, over time it became acceptable in the United States for plum sauce to be served with duck. Eventually—and quite logically—it took on the name duck sauce. Ed Schoenfeld, a restaurateur and Chinese food consultant, says that the Chinese condiment degraded over time as it became mass produced.

Duck sauce is not the only Americanized fare on a Chinese restaurant menu: General Tso's chicken was once very savory, made with garlic and vinegar. Today it is a batter-fried, syrup-laden shadow of its former self. Chop suey and crab rangoon, as well as sweet-and-sour pork, chicken, beef, and shrimp, are other dishes that are Chinese in name only.

For your further enlightenment, here is a list of common ingredients found in duck (plum) sauce: plums, vinegar, sugar, ginger, garlic, chiles, salt, and water. Nope, no duck.

Q How cool are cucumbers?

A In his poem "A New Song of New Similes," eighteenth-century English author John Gay wrote, "I'd be . . . cool as a cucumber could see the rest of womankind." Not at all unwittingly, Gay coined a phrase that has become part of our collective consciousness. But are cucumbers really all that cool?

Let's turn back the clock a few hundred years. Imagine you're in India, the supposed birthplace of the cucumber. It's about one hundred degrees in the shade. You've just polished off a dish of delectable but fiery hot vindaloo chicken, and now it feels like steam is shooting out of your ears and nostrils, like you're a character in a Warner Brothers cartoon. Instinctively, you reach for a glass of ice water. But wait. In your dizzy, sweaty haste, you've forgotten that refrigeration—and Warner Brothers cartoons, for that matter—won't be invented for centuries.

There is no ice water! How do you extinguish the fire that rages in your mouth? That's right: You reach for a few slices of cool cucumber, and they really hit the spot.

While there's no reason to believe that its physical temperature is lower than that of any other vegetable, the cucumber's mild flavor and watery flesh give it a refreshing quality that has made it a favorite warm-weather ingredient in cooling salads, relishes, and yogurt sauces for generations. It should come as no surprise that the cucumber hails from the same family of plants *(Cucurbiticeae)* as the watermelon.

But there is evidence to suggest that cucumbers can keep you "cool" in other ways. The cucumber's skin is a source of fiber, and several studies have shown that a high-fiber diet may help to lower your blood pressure, a benefit for those who share some of the personality traits of certain steam-shooting cartoon characters. And when applied directly to the skin, the ascorbic acid (vitamin C) and caffeic acid found in the cucumber can help soothe and seemingly cool irritations and reduce swelling.

You may have heard of another, non-dietary use for cucumbers. Rock stars have been known to stuff them into their pants to enhance a certain physical feature. The "prop" couldn't be sillier, but they think it makes them look cool.

Q How do they make pink wine if there are no pink grapes?

A We're dedicated to bringing you the most accurate answers to life's important questions. So we spent a lot of time and sacrificed many liver cells at a local vineyard researching this one.

During one particularly lengthy and arduous tasting session, we peppered the vintner with questions about wine production. He answered many of our queries quite pleasantly. The wine flowed. We became brazen. Finally, we asked him to show us his pink grapes. Embarrassment ensued.

It turns out that there are no pink grapes. Pink wines—known to oenophiles as roses—are made from red grapes. Understanding why requires a short lesson in general wine-making. Grapes are put into a crusher, where their juice is extracted and stems and skins are separated from the grape flesh. The flesh and juice of all grapes is basically colorless—wine made from just these elements will be white, regardless of the color of grape skin. This explains why pinot grigio grapes, which can be very dark in color, produce a white wine.

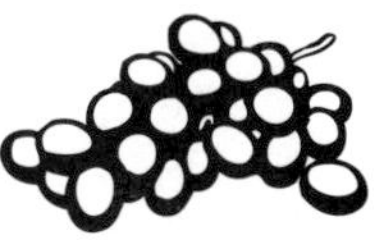

If green grapes are used, the skins are usually removed because green grape skins add little to a white wine. The skins of red grapes, however, bring benefits that include tannins—the element that provides the mouth-puckering feel of dry wines—and an antioxidant called resveratrol. Perhaps most important, including the skins of red and purple grapes in the fermentation process imparts color to the finished product. The amount of contact that is allowed between the grape juice and the grape skins largely determines the color of the wine.

Which brings us back to the pressing question at hand: pink wines. Take white zinfandel (which actually is pink). The zinfandel grape is red. During production, vintners who wish to produce a white zinfandel permit the skins of the grapes to have partial contact with the wine. This tints the wine but doesn't allow it to reach the full red color. Our vintner said that pink wines can also be made by mixing a little red wine with white, although this is rarely done.

And so our inquiry came to a conclusion without revealing any pink grapes. We did, however, eventually see more than a few pink elephants.

Q Which alcoholic beverage gets you drunk the quickest?

A This kind of depends on how you define beverage. The most potent potables aren't exactly the tastiest. But if your palette has a penchant for all things rocket fuel, you have a few options for getting drunk in record time. These include 124-proof absinthe (now legal in some U.S. states), 151-proof Jamaican rum, and 160-proof home-distilled moonshine (this stuff you'll have to get from your friendly backwoods bootlegger—just ask for the "white lightning").

In case no one ever told you how proof works, 160 proof translates to 80 percent alcohol by volume (sometimes abbreviated on liquor bottles as ABV). As far as all of the above spirits go, these are not intoxicants you want to drink straight—add some soda, juice, or iced tea, okay?

Same goes for the winner of our gets-you-loaded-the-quickest contest: Everclear. Everclear, otherwise known as the frat boy's mixer of choice, is a brand of pure grain alcohol that's available in a 190-proof concentration. That's 95 percent alcohol. Many U.S. states have made it illegal to purchase 190-proof Everclear: In those places, you'll likely find a 151-proof version instead. However, the 190 is far and away the firewater of choice, and it has achieved near-cult status. Why? It's pretty darn affordable when you consider the strength-to-dollar ratio (a 750-ml bottle usually sells for between fourteen and seventeen dollars), and it's colorless and odorless, which allows for seamless mixing with your favorite flavor of Kool-Aid or Jell-O.

Speaking of mixing, one of the most popular Everclear concoctions is something called jungle juice. Chopped fresh fruit is soaked in a combination of Everclear, vodka, schnapps, rum, triple sec, wine, and gin for four to twelve hours. Then fruit juice, orange juice concentrate, and clear soda are added to the mix, and the whole brew is allowed to marinate overnight.

Everclear punches like jungle juice are often consumed at parties and, therefore, are often mixed in large containers, such as trash cans. Drink too much, and you'll quite likely end up with your head in one.

Q Are you going to eat that?

A In recent years, Hollywood has learned that people like watching others chow down on calf pancreas and camel spiders. Even so, one person's offal is another person's delicacy. Here are interesting foods enjoyed around the world.

Poi (Hawaii): Taro root is boiled to remove the calcium oxalate poison, then mashed to a muddy purple paste. To most non-Hawaiians, it also tastes like muddy purple paste.

Muktuk (Alaska, Canada): Enjoy a hunk of whale blubber, which looks and feels in the mouth like densely packed cotton soaked in oil. It's attached to a thick piece of whale skin with the look and feel of worn tire tread. You cut off chunks of blubber and chew them—for a long time.

Chorizo (Iberia, Latin America): This is what remains after all the respectable pig offal has been made into normal sausage. By now we're down to the lips, lymph nodes, and salivary glands. It's actually quite spicy and tasty, provided you don't mind dining on an immune system.

Menudo (Mexico): Basically, it's cow-stomach soup. If you can tolerate the slimy, rubbery tripe chunks, the soup itself tastes fine. It's often served for breakfast to cure a hangover.

Scrapple (Pennsylvania): The Amish and Mennonites don't waste much, and pig butchers chop up leftover guts, cook them with cornmeal, then pour it all into bricklike molds to solidify.

Casu marzu (Sardinia): It's illegal in Italy to sell wormy cheese, so pragmatic citizens make their own by sticking a perfectly good round in the cupboard for a couple of months so flies can lay eggs on it. The larvae produce enzymes that break down the cheese into a tangy goo, which Sardinians dive into and enjoy, larvae and all.

Balut (Philippines): Ever get a hankering for soft-boiled duck or chicken embryos? Some Filipinos think there's nothing finer, even though the diner must sometimes pick miniature feathers out of their teeth.

Surströmming (Sweden): Primarily a seasonal dish in northern Sweden, this rotten fermented herring could knock out a wolverine. Even the Swedes rarely open a can of it indoors, except for playful children who swipe some and hide it in their school's air vents.

Jellied eels (England): If you find yourself hungry as you hustle through London, grab a jellied eel from a street vendor. It tastes like pickled herring with a note of vinegar, salt, and pimiento, all packed in gelatin. So next time you're asked to bring something congealed to a potluck, interpret the request loosely and watch the fun that results.

Haggis (Scotland): Drink enough Scotch, and you'll eventually get so hungry you'll eat sheep innards mixed with oatmeal and boiled in the sheep's stomach. Safety-minded haggis chefs suggest poking holes in the stomach so it doesn't explode when the oatmeal expands.

Pombe (East/Central Africa): History shows that people will make alcohol from any ingredients available. That includes bananas, mashed with one's bare feet and buried in a cask. The result is pombe, an east African form of beer.

Durian (Southeast Asia): This football-size fruit with spines poses one of the weirdest contrasts in the culinary world. It smells like unwashed socks but tastes sweet. Imagine eating vanilla pudding while trying not to inhale.

Vegemite (Australia): "Yeast extract" is a brewery by-product that looks like chocolate spread, smells like B vitamins, and tastes overwhelmingly salty. Australians love it on sandwiches or baked in meatloaf, and it goes well with cheese.

Lutefisk (Sweden): It's simply fish boiled in lye. The lye gelatinizes the fish, but if it's soaked too long, the mixture starts turning into soap. The taste is actually fairly mild; the smell depends on the fish used (reportedly, cod isn't the best choice).

Kimchee (Korea): A cultural staple, this spicy dish of cabbage fermented with salt and pepper smells like garbage to many. Most people have less trouble with the taste than the aroma, but we do taste partly with our noses.

Stinkheads (Alaska): If you travel to Alaska's Bering Sea coast someday, stop at a Yup'ik village and ask the natives about their culture. They have great fun introducing visitors to salmon heads that have spent the summer buried in the ground.

Sago beetle grubs (Papua New Guinea): Some tribespeople consider these bugs delicious. Then again, many of the same folks eat a lot of sago pulp (the inside of a palm tree). After months of eating tree innards, perhaps one would relish a roasted bug.

Gulyás (Hungary): Pronounced "guh-yawsh" in Hungarian, this isn't a strange food, but most people would like the real thing better than so-called "goulash," which doesn't do justice to Hungary's national dish. There are probably as many gulyás recipes as there are cooks. The genuine article is a spicy beef-and-potato stew with vast amounts of paprika.

Qat (Horn of Africa): From Yemen to East Africa, people chew this leaf to get a little buzz. One doesn't so much chew it as pack it between his or her cheek and gum to get the full qat pleasure. Bear in mind that it is illegal in the United States.

Ayrag (Mongolia): When you're a nomad of the Gobi and there are no taverns, you're happy to settle for fermented mare's milk. It takes only a couple of days to ferment and turns out lightly carbonated.

Kava (Polynesia): It's the social lubricant of many island nations. Take a pepper shrub root *(piper methysticum)*, and get someone to chew or grind it into a pulp. Mix with water and enjoy.

Balut (Southeast Asia): A favorite in some parts of Southeast Asia, balut consists of a fertilized duck egg that's been steamed at the brink of hatching. Diners, who usually pick up one from a street vendor on the way home from a night of imbibing, chow down on the entire contents of the egg: the head, beak, feet, and innards. Only the eggshell is discarded.

Bird's Nest Soup (China): The key ingredient of this famous dish is the saliva-rich nest of the cave swiftlet, a swallow that lives on cave walls in Southeast Asia. This delicacy is so popular that it has endangered the bird's population. Some enterprising suppliers have started farming nests by providing birds with houses in which to build. Still, wild nests remain the most highly prized.

Century Eggs (China): When Sam I Am was expressing his disgust for green eggs in Dr. Seuss's classic book, he was probably talking about the Chinese specialty known as century eggs, also called thousand year eggs or preserved eggs. To make the dish, chicken, duck, or goose eggs are preserved in a mixture of salt, lime, clay, ash, and rice husks or tea leaves, then allowed to ferment for several weeks or months. The process turns the egg whites into gelatinous, transparent dark-brown masses, while the yolks become pungent yet intensely flavored orbs of variegated shades of green. What's not to like?

Cobra's Blood Cocktails (Indonesia): Those looking for an exotic drink in Indonesia can quench their thirst with a fresh cocktail of cobra's blood, served either straight or mixed with liquor. Depending on the establishment, consumers can choose between different types of cobras to supply the blood, with the King Cobra being the most prized and expensive.

Kopi Luwak Coffee (Indonesia): Java junkies may need to save up to indulge in this exotic coffee. What makes it so special? For kopi luwak coffee, only the sweetest, tastiest coffee beans are picked by experts—not human experts, but rather wild civets who live in Indonesia's coffee-producing forests. The animals nibble on the fruity exterior before swallowing the hard inner beans. During the digestive process, their gastric juices remove some of the proteins that ordinarily make coffee bitter. Humans don't intervene in the coffee-making process until it's time to separate the undigested beans from the civet dung. Experts claim the result is an exceptionally smooth and balanced cup of Joe.

Seal Flipper Pie (Newfoundland): This maritime favorite is made from the chewy cartilage-rich flippers of seals, usually cooked in fatback with root vegetables and sealed in a flaky pastry crust or topped with dumplings.

HOLIDAYS AND CELEBRATIONS

Q Why do we drink eggnog at Christmas?

A Nothing says Christmas quite like the mixture of raw eggs, hard liquor, and fragrant spice that we call eggnog. What could be more festive than this dubiously textured, high-alcohol punch? Sure, there are plenty of ways to get plastered during the holiday season, but only eggnog can incline the drinker toward that unholy trinity of drunkenness, salmonella poisoning, and heart disease.

But it's not just about getting wasted. You can buy non-alcoholic eggnog—heavily processed and artificially flavored—at any grocery store. It might be because the holiday season has traditionally been a time of rich foods and sweet desserts—and isn't a glass of eggnog basically a Christmas cookie in liquid form?

This supermarket concoction is the most modern example of a venerable (and patriotic) tradition. Eggnog dates back to colonial America. George Washington is said to have been so fond of the drink that he developed his own recipe, which included whiskey, rum, brandy, and sherry. (That'll warm your Valley Forge!) George wasn't the only Virginia farmer-politician who enjoyed the tipple. Another story has it that in the early days of the

Virginia House of Delegates, a Christmas-season session had to be adjourned because the members were so loaded on eggnog that they were in a "helpless condition."

There are some Scrooges who claim that eggnog isn't really an American innovation at all. And it's true that our founding fathers weren't the first to mix eggs and booze to make a delicious drink. There are earlier English punches that combine egg yolks with wine or beer—one explanation of the origin of the word "eggnog" claims that "nog" is a reference to a specific type of English beer. And some reckless speculators have guessed that our beloved beverage was inspired by a French recipe. *Mon Dieu!*

In all fairness, let's give credit where credit is due. Even if we didn't think up the idea, we gave it a catchy name, made it stronger, and drank the hell out of it. And after all, isn't that what Christmas is all about?

Q How did high school proms get started?

A The word itself seems to have come from slang-happy college students of the late eighteen hundreds and early nineteen hundreds. It was originally a shortened form of "promenade," meaning the entry and announcement of guests at a formal dance, and then became a term for college dances in general. The earliest known reference to a prom can be found in an Amherst College student's 1894 journal entry; it likely referring to a senior class dance held at nearby Smith College.

But what about high school dances—proms, specifically? They are more closely related to upper-class debutante balls than to college dances. The debutante ball originated in sixteenth-century

England as a formal way to present a young woman who was available for marriage. The tradition spread to America in the late nineteenth century and flourished among the wealthy as a rite of passage to adulthood. In the early twentieth century, middle-class parents wanted the opportunity to give their own kids similar tastes of adulthood, so they began organizing high school dances. References to these proms started popping up in high school yearbooks in the 1930s, but some communities were likely holding these events earlier than that.

Early proms were far cries from the extravaganzas of today. High school kids would dress up in their Sunday best, go to the school gym, enjoy some refreshments, and cut a rug—all under the watchful eyes of adults. Limousines and tuxedos weren't yet part of the equation.

But as America's middle class grew more prosperous in the 1950s, kids and parents spent more money on proms. In turn, the importance of these dances increased. Some schools moved them out of their gyms and into hotel ballrooms; young men started cramming themselves into formal wear, and young ladies began dropping loads of cash on special prom dresses. The popularity of proms waned a bit in the 1960s, when everybody who was anybody rebelled against everything. But proms came back strong in the mid-1980s, fueled by a popular string of romantic teen comedies, such as *Footloose* and *Pretty in Pink.*

These days, proms are big business. Some estimates place total annual U.S. prom expenditures at more than two billion dollars. In 2007, *Seventeen* magazine estimated that the girl alone drops an average of eight hundred dollars on her senior prom. That's a lot of money to spend for a date with a guy you'll probably never see again after graduation.

Q Who dreamed up the bachelor party?

A Maybe you're a conspiracy theorist, and you think that a young bride-to-be came up with the idea as a way of getting the upper hand before the marriage was even consummated. It would be a clever ploy, setting up the groom for trouble and forcing him into months of repentance right off the bat.

But in his book *Bachelor Party Confidential: A Real-Life Peek Behind the Closed-Door Tradition,* David Boyer traces the dawn of the bachelor party all the way back to about 500 B.C., when soldiers in ancient Sparta would gather for lavish dinners, during which the guests exchanged toasts and the groom-to-be reiterated his commitment to his fellow men.

Boyer says that it's difficult to pin down exactly when the racier elements of today's bachelor parties began to creep into the festivities, but he notes that a high-society bachelor bash in New York City in 1896 was raided by the police, who had reason to believe that naked women were part of the entertainment. They were wrong—it was just a belly dancer. Nevertheless, the saucy shindig made headlines. The fact that the police had their suspicions suggests that even in those days, rowdies were getting away with the types of wild bachelor parties we've come to know and love today.

Boyer is the Studs Terkel of stag, relying on oral histories to show how the bachelor party evolved over the course of the twentieth century. While the parties were wild enough in the pre-World War II days, they were primarily the domain of the upper class. After the war, however, when returning servicemen were getting married en masse, the modern bachelor party—with its requisite booze, strippers, and various forms of insanity—became a true cultural phenomenon.

So the next time you wake up at sunrise in a pool of your own vomit after celebrating a buddy's impending marriage, take solace in the fact that you've merely been carrying on a noble tradition that was started by the Spartan soldiers.

Q Why is April 1 such a foolish day?

A Search all you want for an explanation of the origin of April Fools' Day—eventually you'll discover that you're on a fool's errand. You can find quite a few attempts to explain why people play pranks on each other every year on April 1, but none are definitive. In fact, some of those stories are themselves April Fool's jokes.

In 1983, The Associated Press published an explanation provided by Professor Joseph Boskin of Boston University. Boskin said that the Roman emperor Constantine I once allowed a court jester to rule as king for a day on April 1, and the jester issued a proclamation that called for absurdity on that day, triggering an annual tradition. The story is complete fiction: Boskin was playing his own April Fools' joke on the AP reporter.

Nobody knows for sure why, when, or how April Fools' Day began. Some peg the tradition to the switch from the Julian calendar to the Gregorian calendar during the sixteenth century, mandated by Pope Gregory XIII. The old Julian calendar, as observed in Christian Europe, placed the start of the new year on March 25, and in many Christian countries, New Year's festivals took place from March 25 to April 1. It is said that after the new calendar was put into use, people who continued to celebrate the new year on the old date came to be known as "April fools." Many historians, however, believe that the tradition probably emerged from the general foolishness and frivolity associated with spring festivals that usually took place in late March and early April.

The day takes on different forms in different places. In England, it's referred to as All Fools' Day. In Scotland, an April Fool is referred to as a "gowk," a Scottish term for "cuckoo." In France, the victim of a prank played on April 1 is called *poisson d'Avril,* or "April fish." Often, the actual prank involves attaching a picture of a fish to the unsuspecting victim's back.

Those wacky French. How could we laugh without them?

Q What is Dyngus Day?

A When was the last time you were awoken by the icy-cold sensation of a bucket of water being poured over your head? If you're among the thousands of folks who celebrate Dyngus Day, the answer is the day after Easter.

Dyngus Day (also called Dingus Day, Easter Monday, Smigus Dyngus, or Wet Monday) is a Polish holiday that originated around A.D. 966, when Poland's Prince Mieszko I accepted Christianity and was baptized, along with his entire court. Since then, the celebration has evolved from an annual mock-baptism to a sort of courting ritual, during which a young man douses the girl of his dreams in hope that she'll be flattered.

Yes, you read that right. It seems like a strange tradition, but it's one still practiced today, especially in the communities of Buffalo, New York, and South Bend, Indiana. In those cities, everyone is packing at least a water pistol on Dyngus Day, and some more enterprising soakers make use of garden or even fire hoses. Originally, boys were the only ones armed with buckets of water, but in recent decades, the girls have begun to fight back, launching their own H2O-based assaults.

Traditionally, Dyngus Day has meant more than just a water fight. In addition to the hydro-powered alarm clock, boys would fashion small whips out of pussy willow or birch branches, and use these to strike their paramours on the shins. In Poland, where matchmaking is a very big deal, a young girl who didn't receive these attentions was considered hopeless, romantically speaking.

Mercifully, the shin-swatting tradition has largely fallen by the wayside, and participants in Dyngus Day now focus almost exclusively on the irreverent fun that goes along with a citywide water war. Visitors to Poland, Buffalo, or South Bend on the day after Easter are advised to bring a few changes of clothes, and perhaps a bandolier of water balloons.

Q Why do we put candles on a birthday cake?

A Are we feeling a little sensitive about the five-alarm fire that is blazing atop the buttercream frosting? Well, take heart—the tradition of lighting candles on cakes is way older than you. The custom dates back to the ancient Greeks.

It all began as an offering to Artemis, goddess of the moon. The Greeks baked round honey cakes, topped them with tapers, and placed them on the altar of Artemis's temple. When lit, the round cakes looked like—you guessed it—full, glowing moons. Back then, people believed that smoke carried their thoughts up to the gods (hence, all of the sacrificial fires). These days, we associate lighting and blowing out the candles with making a wish. But just when did candle-topped cakes become an essential part of the party?

Many historians trace the modern use of candles on cakes to Kinderfest, a German birthday celebration for children that dates to the fifteenth century. In those days, people believed that children were particularly susceptible to evil spirits on their birthdays, so friends and family gathered around protectively, lighting candles on a cake to carry good wishes up to God. It was customary for the candles to remain lit all day, and the cake was served after the evening meal.

By the eighteenth century, birthday cakes and candles took on a more festive feel. The research of culinary historian Shirley Cherkasky points to a 1799 letter written by Johann Wolfgang von Goethe (one of the greatest figures of world literature) that recounts his fiftieth birthday: "When it was time for dessert, the prince's entire livery in full regalia entered, led by the majordomo. He carried a generous-size torte with colorful flaming candles—amounting to some fifty candles—that began to melt and threatened to burn down, instead of there being enough room for candles indicating upcoming years."

Well, what do you know? You and Goethe actually have something in common: birthday cakes that resemble towering infernos. But, hey, controlled fires can be really fun (just ask any pyromaniac).

And perhaps that's why the German birthday cake tradition eventually made its way over to the United States. As an 1889 American style guide directed: "At birthday parties, the birthday cake, with as many tiny colored candles set about its edge as the child is years old, is, of course, of special importance."

By 1921, American candle manufacturers started advertising boxes of little candles in mixed colors. And a few years later, people all over the United States could order cake candles and candle holders from the famous Sears Roebuck catalog. But you're way too young to remember that. Right?

Q How did Groundhog Day get started?

A It wasn't the brainchild of Punxsutawney Phil, the world's most famous weather-predicting groundhog. February 2, the day we observe as Groundhog Day, is important in the seasonal cycle. It falls halfway between the winter solstice—the shortest day of the year in the Northern Hemisphere—and the spring equinox, which is one of two days of equal sunlight and darkness.

This midpoint between winter and spring brings anticipation of a weather change from harsh cold to pleasant warmth. If we humans have an opportunity to believe that warm weather may come sooner rather than later, we'll take it. So weather-predicting became tied to this day.

The ancient Celts marked this halfway point with a holiday called Imbolc. Early Christians, meanwhile, routinely scheduled holidays to compete with and replace pagan holidays. February 2 is forty days after December 25, so Imbolc became Candlemas Day, an observance of the day Mary and Joseph took Jesus to the temple to perform the redemption of the firstborn. It was celebrated as a sort of end to the Christmas season.

The weather-predicting aspect of Candlemas Day carried over from pagan traditions. Europeans would place a candle in their windows on the eve of Candlemas Day. If the sun was out the next morning, they believed it indicated that there would be six more weeks of winter.

The predicting tools varied across Europe; in some places, they involved animals, such as bears, badgers, and, in Germany, the hedgehog. But requirements were the same: If it was sunny and the animal cast a shadow, it meant a longer winter; if it was cloudy and there was no shadow, it meant a shorter winter.

When the hedgehog-watching Germans came to hedgehog-free America, they became groundhog-watching Americans. The earliest reference to Groundhog Day in America can be found in an 1841 diary by James Morris, a Pennsylvania shop owner.

With such a proud legacy, groundhogs can be counted on to be correct most of the time, right? Think again. A Canadian study showed that groundhogs are correct in predicting the length of winter only about 37 percent of the time.

Q Why do we toast by clinking glasses?

A There was nothing like a few libations to get the questions flowing for this book. Unfortunately, those questions usually wound up being something along the lines of, "Whose pants are these, and why am I wearing them on my head?" But on one occasion—after our tenth or eleventh toast—we wondered just why we were clinking our glasses together. Hey, a legitimate question!

As is often the case with matters such as this, there is the interesting explanation and then there is the real one. For many years, the dominant theory held that clinking glasses after a toast was not a sign of friendship, but of mistrust. Back in ancient times, when poisoning was the preferred method of dispatching one's enemies, a host would clink glasses with his guests, making sure they sloshed a little vino into each other's cups to prove that the wine was not poisoned. This theory has been largely discredited.

Besides the logistical nightmare (not to mention the cleaning bill) of trying to get wine into other people's cups without spilling, it seems unlikely that ancient social events were so perpetually treacherous that this would evolve into a common social custom.

To understand why we clink glasses after a toast, we need to examine where the custom of toasting began. It does seem a little strange, after all—we don't thrust our food-laden forks into the air and offer syrupy platitudes before taking the first bite of a meal. Then again, in the Western tradition, eating from the same plate doesn't have quite the same history that drinking from the same cup does.

Indeed, many historians identify toasting's origins in the drinking habits of the ancient Greeks. In those heady, wine-filled days, it was customary for groups of acquaintances to share the same drinking vessel. The host would take the first sip and then pass the cup to the next person in the party, often accompanying this gesture with a pleasantry or a wish for good health. As the centuries passed, drinking, though still a largely communal act, no longer involved a collective drinking vessel. So clinking became popular, to provide the communal sense once afforded by a shared drinking vessel.

Why do we call it a "toast"? In medieval British times, people drank whatever they could get their hands on. The Brits, never known for their culinary expertise, didn't produce the finest wines and spirits, and to make their sometimes acidic quaffs more palatable, on occasion they would drop a piece of spiced, charred bread into the drink. (Charcoal has anti-acidic properties.) Shakespeare fans will surely remember Falstaff, possibly the most glorious drunk of all time, demanding a "quart of sack" with "toast in't" in *The Merry Wives of Windsor.*

There is another interesting, if questionable, theory about toasting that bears mentioning. It's been suggested that clinking glasses derives from the medieval belief that alcohol was possessed by demons. Medieval folks thought that loud noises would drive demons away, and so a clinking of the glasses prior to imbibing would inhibit the evil spirits from making you do bad things. This theory has been discredited, but the idea of evil spirits inhabiting alcohol has continued to the present day. Indeed, we've tried this excuse ourselves. Trust us, it doesn't work.

Q What is Watch Night?

A In one sense, New Year's Eve may be the most universal of holidays, as virtually every culture on the planet has some way of marking the passing of one era and the start of another. Of course, each culture celebrates at its own time of year and with its own set of traditions. For most Americans, December 31 is a time for counting down and drinking up, and January 1 is a time of empty resolutions and epic hangovers. But even in America there are alternative ways to mark the passing of an old year and the coming of a new one.

For many black Americans, the New Year's holiday centers around one of the most meaningful and popular religious services of the year: Watch Night. Throngs of people gather in churches on New Year's Eve for a long evening of song, reflection, celebration, and prayer in a ceremony that can fluctuate between pious and raucous.

Watch Night originated with the Moravians, Christians whose roots trace back to the present-day Czech Republic. The first such services likely took place in Germany in 1733. In 1777, Methodism founder John Wesley brought the practice to the states, calling for his followers to renew their covenant with God in a monthly vigil that corresponded to the full moon.

It is commonly believed that the tradition of celebrating Watch Night on New Year's Eve dates to the American South during the days of slavery. This much is true. But why? Some sources claim that slave families gathered to pray together on New Year's Eve because they feared it would be their last night together. Many slaves saw their families broken up after New Year's when individuals were sold to raise funds to settle outstanding debts. But this explanation falls short. The most significant evidence against this theory is that Southern businessmen were unlikely to conduct business on New Year's Day.

What is known is that Watch Night became an important African American celebration during the Civil War, as slaves and free blacks across the country gathered in their churches on New Year's Eve 1862 to await a watershed moment in United States history. The next day, January 1, 1863, President Abraham Lincoln was scheduled to—and did—sign the Emancipation Proclamation. This simple but eloquent 725-word document freed everyone held in slavery in the Confederate States and declared all of them eligible for military service in the Union. Then and now, the document was criticized for having freed only those slaves residing in states that were in rebellion, but it irrevocably committed the federal government to abolishing slavery throughout the entire country.

The Watch Night tradition continued in black congregations for decades as a celebration of hard-won freedom and then waned in the early years of the 20th century. The practice was reinvigorated in the 1960s as part of the Civil Rights Movement and continues to this day, serving as a powerful reminder and celebration of one of the most pivotal moments in American history. Celebrations vary from church to church, but there are many commonalities. Services typically begin late, around 10 PM, and continue until shortly after midnight, though some groups stay throughout the night and celebrate until dawn. One congregation will often host another, so that members of both churches can share the night together. There's usually a sermon, or the pastor may read the official church record from the year that's ending. Members of the church will often give spontaneous testimonials of the trials they've faced and the blessings they've received, putting the former behind them and expressing gratitude for the latter. The service sometimes includes a candlelight procession and is always interspersed with joyful singing. Just before midnight, worshippers crowd together around the altar and fall to their knees to pray in a moving scene that dedicates the start of the year to God. And so continues the Methodist notion of renewing one's covenant at New Year's.

Q How did Easter eggs come to be?

A Easter is both the most solemn and the most joyous event on the Christian calendar. But the history of this celebration is complicated, and modern Easter traditions include elements from a variety of spiritual practices. For instance, the term paschal describes the Christian season but also refers to Pesach, the Jewish Passover.

Decorated eggs have long been a part of Easter observances. In the 13th century, the church didn't allow eggs to be eaten during

Holy Week, the week prior to Easter Sunday. Eggs reappeared on the dinner table on Easter Day as part of the celebration. These eggs were often colored red to symbolize joy. Hunting for eggs and giving decorated eggs as gifts came later.

Why, though, does the Easter Bunny deliver the eggs? When did this tradition begin? The link between Easter and rabbits is ancient. The word Easter is likely derived from the name of the mother goddess Eostre, who was worshipped by the ancient Saxons of northern Europe. Her festival coincided with the lengthening days that marked the arrival of spring and the return of life after a barren winter. Eostre's emblem was the hare; rabbits and hares have often been regarded as symbols of fertility. In one legend, Eostre magically changed a beautiful bird into a hare that built a nest and laid eggs.

Myths and folk traditions frequently offer explanations for natural events. Hares raise their babies, which are called leverets, in "forms." A form is a hollow in the ground of a field or meadow. Female hares, or does, often divide a litter among two or three forms for safety. Abandoned and empty forms attract plovers, a kind of wading bird, which occasionally move in and use the form as a nest for their own eggs. People saw hares in fields—seemingly hopping away from forms full of eggs—and concluded that the hares had laid the eggs.

As Christianity spread throughout the known world, it absorbed pagan beliefs and practices, endowing them with Christian meaning. Church officials placed observances of the events surrounding the crucifixion of Jesus in the early spring. Eggs, the source of new life, came to stand for Christ's resurrection from the tomb. The celebration of spring and the worship of the goddess Eostre became Christian Easter. In time, as distinctions between hares and rabbits were blurred, Eostre's hare became the Easter Rabbit.

The link between rabbits and Easter emerged most strongly in Protestant Europe during the 17th century, particularly in Germany. Boys and girls built "nests" with their caps and bonnets, and good children were rewarded with a "nest" full of colored eggs brought to them by the *Osterhas*, or Easter Rabbit. Variations of this practice came to America in the 18th century, especially with immigrants from Germany. A hunt for decorated Easter eggs left by the Easter Bunny became common in the 19th century.

Q Who owns "Happy Birthday"?

A Many people believe that "Happy Birthday to You"—the most frequently sung song in the English language—is a traditional folk melody that rests comfortably in the public domain. In fact, the song is protected by strict copyright laws.

That four-line ditty is as synonymous with birthday celebrations as a cake full of candles. According to the *Guinness Book of World Records*, the most popular song in the English language is "Happy Birthday to You." What is less well known, however, is that it is not a simple tune in the public domain, free for the singing by anyone who chooses. It's actually protected by a stringent copyright that is owned and actively enforced by the media conglomerate AOL Time Warner. You are legally safe to sing the song at home, but doing so in public is technically a breach of copyright, unless you have obtained a license from the copyright holder or the American Society of Composers, Authors, and Publishers.

The popular seven-note melody was penned in 1893 by two sisters, Mildred J. Hill and Patty Smith Hill, as a song titled "Good

Morning to All." It remains unclear who revised the words, but a third Hill sister, Jessica, secured copyright to the song in 1935. This copyright should have expired in 1991, but through a number of revisions to copyright law, it has been extended until at least 2030 and now lies in the hands of AOL Time Warner. The company earns more than $2 million a year from the song, primarily for its use in movies and TV shows. Because licensing the rights to the song is a costly endeavor, low-budget movies have to cut around birthday scenes, and many popular chain restaurants insist that their employees sing alternate songs to celebrate their customers' birthdays.

Unless you license the rights, singing the song in public could result in something decidedly unhappy. And if you're going to be arrested on your birthday, don't you want it to be for something more exciting than copyright infringement?

Q Which came first, the turkey or Thanksgiving?

A Most people were taught that Thanksgiving originated with the Pilgrims when they invited local Native Americans to celebrate the first successful harvest. Here's what really happened.

There are only two original accounts of the event we think of as the first Thanksgiving, both very brief. In the fall of 1621, the Pilgrims, having barely survived their first arduous year, managed to bring in a modest harvest. They celebrated with a traditional English harvest feast that included food, dancing, and games. The local Wampanoag Indians were there, and both groups demonstrated their skill at musketry and archery.

So that was the first Thanksgiving, right? Not exactly. To the Pilgrims, a thanksgiving day was a special religious holiday that consisted of prayer, fasting, and praise—not at all like the party atmosphere that accompanied a harvest feast.

Our modern Thanksgiving, which combines the concepts of harvest feast and a day of thanksgiving, is actually a 19th-century development. In the decades after the Pilgrims, national days of thanksgiving were decreed on various occasions, and some states celebrated a Thanksgiving holiday annually. But there was no recurring national holiday until 1863, during Lincoln's administration, when a woman named Sarah Josepha Hale launched a campaign for an annual celebration that would "greatly aid and strengthen public harmony of feeling."

Such sentiments were sorely needed in a nation torn apart by the Civil War. So, in the aftermath of the bloody Battle of Gettysburg, President Lincoln decreed a national day of thanksgiving that would fall on the last Thursday in November, probably to coincide with the anniversary of the Pilgrims' landing at Plymouth. The date was later shifted to the third Thursday in November, simply to give retailers a longer Christmas shopping season.

So what about the turkey? Does that date back to Pilgrim times? Well, maybe. Governor William Bradley's journal from around that time indicates that "besides waterfowl there was great store of wild turkeys, of which they took many." Another record notes that "our governor sent four men on fowling…they four in one day killed as much fowl, as with a little help beside, served the company almost a week."

Of course, "fowl" doesn't necessarily mean turkey, so the best we can say is that the Pilgrims may have eaten it. The only food we know for certain they ate was venison, and that was provided by their guests, the Native Americans (who may have been a little surprised by the meager spread their hosts had laid out). They probably also ate codfish, goose, and lobster, but not a lot of vegetables—you can catch fish and fowl, but it takes time to grow crops. As for that Thanksgiving stape mashed potatoes? Nope—potatoes hadn't yet been introduced to New England.

So how did the gobbler become the centerpiece of Thanksgiving celebrations? It may have had something to do with the prevalent diet at the time the national holiday was founded in 1863. Beef and chicken were too expensive to serve to a crowd, and even if you had your own farm, you needed the animals' continuous supply of milk and eggs. Venison was an option, but you couldn't always count on bagging a deer in time for the holiday. Turkey was readily available, not too expensive—and very popular, perhaps in part due to the scene at the end of Charles Dickens's *A Christmas Carol* in which Scrooge buys "the prize turkey" for Bob Cratchit's family. The novel, published in 1843, was immensely popular in America and may have secured the humble fowl's center-stage spot on the Thanksgiving table for generations to come.

Incidentally, according to USDA statistics, between 260 million and 300 million turkeys are consumed each year in the United States. As well as being the long-standing centerpiece at holiday dinners, roast turkey was one of the meals that Neil Armstrong and Buzz Aldrin ate on the moon (it was a "wetpack" dinner that included cranberry sauce and stuffing, heated with the assistance of hot water). For a bird that's unable to fly, the domesticated turkey is certainly well traveled.

Q Aren't there better ways for Santa to sneak into a house than crawling down a chimney?

A Except for a few sour souls, everyone loves Santa Claus. How could you not? He spends his days in an enchanted world of elves and toys, he has an awesome flying sleigh, and he has the godlike ability to watch all the world's children at the same time. Yet for somebody who is supposed to be such a magical, all-knowing being, it appears that Santa Claus possibly isn't very bright. Case in point: this chimney business.

Come on, Claus. Do you really need to shimmy down a filthy chimney to deliver your presents? And with that ridiculous diet of cookies and milk, how much longer will you fit down it?

In Santa's defense, there's a lot of tradition behind his chimney act. Even though the image of the red velvet–clad Santa known to most Americans is a fairly recent development, the figure of Father Christmas is rooted in traditions dating back centuries. And though most people know that the Christian figure of Santa Claus is loosely based on Saint Nicholas from the fourth century—Saint Nick is one of Santa's nicknames, after all—most of Santa's behavior, and magical powers are drawn from pagan sources.

Indeed, historians claim that not only Santa Claus, but also much of the holiday of Christmas itself is rooted in pagan tradition. Back in pre-Christian Europe, Germanic people celebrated the winter solstice at the end of December with a holiday known as Yule. Christmas, which later supplanted the pagan winter solstice festivals during the Christianization of Germanic people, maintained many of the pagan traditions. One was the belief that at Yule-time, the god Odin would ride a magical eight-legged horse through the sky. Children left food for the horse, which would be replaced by gifts from Odin, a custom that lives on today in the form of cookie bribery for Kris Kringle and his flying reindeer.

As for sliding down the chimney, folklorists point to another Germanic god: Hertha, the goddess of the home. In ancient pagan days, families gathered around the hearth during the winter solstice. A fire was made of evergreens, and the smoke beckoned Hertha, who entered the home through the chimney to grant winter solstice wishes.

It wasn't until 1822, when literature professor Clement Clarke Moore penned "Twas the Night Before Christmas," that Santa sliding down the chimney became a permanent fixture in popular Christmas tradition. Moore's poem became even more influential forty years later, when legendary cartoonist Thomas Nash illustrated it for *Harper's* magazine. In Nast's depiction, Santa was transformed from the skinny, somewhat creepy-looking figure of earlier traditions into a jolly, well-bearded soul. Despite Santa's physical transformation, other traits from his early incarnations linger, including the bewildering habit of crawling down chimneys.

But just because something is a habit doesn't make it excusable. The figure of Santa has morphed over the centuries, and there's no reason why he can't break the chimney routine in the future. Let's go, Santa—it's time to join the twenty-first century.

Q Is Festivus a real holiday?

A Frank Costanza—George's father in *Seinfeld*—isn't exactly a grinch, but in one classic episode he is out to change the holiday season as we know it. As he explains it, the experience of fighting with another father over a toy doll at Christmas years ago led him to seek an alternative to Christmas.

"Out of that," he says, "a new holiday was born: a Festivus for the rest of us!"

According to Frank, the made-up holiday is commemorated with three primary traditions: airing grievances with family members, performing "feats of strength," and gathering around the Festivus pole (a bare aluminum pole).

Festivus originated with *Seinfeld* writer Daniel O'Keefe, whose family first celebrated their own holiday called "Festivus" in February 1966, to commemorate the anniversary of O'Keefe's parents' first date. While the Festivus pole was not part of the original celebration, the 1966 version did include a wrestling match between the children and audio recordings in which each family member described how the others had disappointed them over the previous year.

Seinfeld fans embraced Festivus with glee, and it quickly entered the pop culture lexicon. Festivus celebrations, both public and private, caught on. One entrepreneurial business, The Wagner Companies, even began manufacturing Festivus poles. In 2005, Wisconsin governor Jim Doyle displayed a Festivus pole in the governor's residence. Two years later, the mayor of Green Bay rejected a resident's request to display a Festivus pole at city hall—even after the city had decided to display a Wiccan pentacle in addition to a nativity scene as a nod to religious inclusivism. The mayor called the Festivus pole request "silly antics"—which is really what Festivus is all about.

The father of writer O'Keefe (also named Daniel), the founder of Festivus, told *The New York Times* that the name of the holiday "just popped into my head." The English word festive comes from the Latin festum, meaning "feast." English also has festal, as in "festal atmosphere" or "festal banners," but the adjective now sounds archaic. And in case you were wondering, yes, the plural of Festivus is Festivi.

Q Why do people wear wedding rings?

A A wedding ring, traditionally a simple gold band, is a powerful symbol. The circle of the ring represents eternity and is an emblem of lasting love in many of the world's cultures. But the history of the wedding ring also includes less spiritual associations.

Historians suggest that wedding rings are a modern version of the ropes with which primitive men bound women they had captured. This suggests that the phrase "old ball and chain" may have referred more to the passage of a bride from person to prisoner than to the husband's matrimonial outlook.

Our current perceptions of the wedding ring evolved over time and are rooted in a variety of ancient practices. In Egypt, a man placed a piece of ring-money—metal rings used to purchase things—on his bride's hand to show that he had endowed her with his wealth. In ancient Rome, rings made of various metals communicated a variety of political and social messages. In marriage, the ring holding the household keys was presented to the wife after the ceremony when she crossed the threshold of her new home. Later, this key ring dwindled in size to a symbolic ring placed on the woman's finger during the wedding. Among Celtic tribes, a ring may have indicated sexual availability. A woman might have given a man a ring to show her desire; putting her finger through the ring may have symbolized the sexual act.

Wedding rings were not always made of gold. They could be made of any metal as well as leather or rushes. In fact, in the 13th century, a bishop of Salisbury in England warned young men against seducing gullible virgins by braiding rings out of rushes and placing them on their fingers. In the 17th century, Puritans decried the use of a wedding ring due to its pagan associations and ostentatious value, calling it "a relique of popery and a diabolical circle for the devil to dance in."

Why is the ring placed on the fourth finger of the left hand? According to fourth-century A.D. Roman grammarian and philosopher Ambrosius Theodosius Macrobius, the fourth finger is the one most appropriate to that function. Macrobius described the thumb as "too busy to be set apart" and said "the forefinger and little finger are only half-protected." The middle finger, or medicusm, is commonly used for offensive communications and so could not be used for this purpose. This leaves only the fourth finger for the wedding ring.

It was once believed that a vein ran directly from that finger to the heart, the so-called *vena amoris*, or vein of love. Since the right hand is commonly the dominant hand, some scholars suggest that wearing a ring on the left, or "submissive," hand symbolizes a wife's obedience to her husband. The importance of the Trinity in Christian theology provides another explanation for the identification of the fourth finger on the left hand as the "wedding ring finger." In the early Catholic Church, the groom touched the thumb and first two fingers of his bride as he said, "In the name of the Father, Son, and Holy Ghost." He then slipped the ring all the way onto the next, or fourth, finger as he said, "Amen." This four-step placement of the ring was a common custom in England until the end of the 16th century, and it remained a tradition among Roman Catholics for many more years.

Q How are weddings celebrated around the world?

A Different cultures celebrate the bride and groom in some pretty fascinating ways. Let's look at some of them.

"Handfasting" is a tying ritual practiced in one way or another throughout the world. In some African tribes, it involves tying together the wrists of the bride and groom with cloth or braided grass during the wedding ceremony. For Hindus, a string is used, and for the ancient Celts, handfasting was the complete wedding ceremony: A year and a day after the tying ritual, the couple was legally married.

Among the *fellahin* in northern Egypt, the priest conducts the handfasting ceremony by tying silk cord over the groom's right shoulder and under his left arm; then he says a prayer and unties him. Next, the priest ties the wedding rings together with the same cord, and after questioning the bride and groom about their intentions, he unties the rings and places them on the couple's fingers.

Though they don't call it "handfasting," Thai couples link their hands together for the wedding ceremony with a chain of flowers, while Laotians use a simple white cotton string. But why just tie the couple's hands together when you can tie up their whole bodies? Guatemalan couples are "lassoed" together with a silver rope; Mexican couples with a white rope or rosary. In a traditional Scottish wedding, the bride and groom tie strips of their wedding tartans together to symbolize the union of their two clans.

Shattering crockery for good luck is a "smash hit" in a number of cultures. Russians throw their champagne glasses on the ground, as do the Greeks (along with their plates). Jewish weddings end with the breaking of a wine glass to symbolize one of three things:

the destruction of the ancient Temple of Jerusalem, the end of the bride and groom's past lives, or that the couple will share as many years as there are shards of glass.

Italian couples also count the shards from a broken glass or vase to see how many years they'll be happily married. Ukrainians follow a tradition called Vatana, breaking dishes with silver dollars to symbolize future prosperity, while the German custom is to host a pre-wedding dish-smashing party, called the Polterabend, during which family and friends shatter china (because glass is considered bad luck) for the engaged couple to clean up—the first of many messes they'll have to deal with as husband and wife. Bulgarian brides raise the stakes by filling the dish with food—wheat, corns, and raw egg—before tossing it over their heads; an English bride might drop a plate of wedding cake from her roof.

Other cultures skip the dish and just break the food. Hungarian brides smash eggs to ensure the health of their future children, and Sudanese ceremonies are marked by the breaking of an egg outside the couple's new home to symbolize the groom's role as master of the house. Many Middle Eastern cultures observe a pre-wedding "grinding" ritual in which the unmarried girls drape a cloth over the heads of the bride and groom and one of the girls—the "grinding girl"—grinds two lumps of sugar over them to repel evil spirits. The Iranian twist on this ceremony involves shaving crumbs from two decorated sugar cones over the heads of the newlyweds for luck.

Circling is also a custom practiced in many places. Hindu couples finalize their union by taking seven steps around a ceremonial fire. Seven is also the magic number for Jewish couples. Traditionally, after stepping under the chuppah, or wedding canopy, the bride circles the groom seven times to represent the seven wedding blessings and seven days of Creation—and also to demonstrate her subservience to the groom. (In modern ceremonies, the bride and groom will often circle each other to show equality.)

For other cultures, three is the lucky number for circles. In the Eastern Orthodox tradition, a priest leads the couple in their first steps as husband and wife three times around the altar—to signify the dance around the Ark of the Covenant—while the choir sings three ceremonial hymns. Croatian wedding guests circle a well three times in honor of the holy Trinity and toss apples into it to ensure the couple's fertility. Moroccan brides circle their new home three times before entering it and officially assuming the role of wife. So if you're engaged, think handfasting, smashing, and circling, and you'll be in good global company.

LOST (AND SOMETIMES FOUND)

Q How come ruins are always underground?

A Ruins actually aren't always underground, but do you think that's going to stop us? We're here to focus on those that are buried. Natural disasters are an interesting place to start. If an earthquake hit or a volcano erupted, an ancient site could be decimated and buried pretty easily, which is what happened to the city of Pompeii.

In Egypt, the powerful desert sands could quite easily have buried something as large as an ancient city. For example, the ancient temple of Abu Simbel was hidden for thousands of years, until Giovanni Belzoni dug into the entrance in 1817.

Many sites around the world were abandoned, for one reason or another, and became overgrown with vegetation. When the vegetation rotted, a layer of soil was formed, and the sites' journeys underground began. Furthermore, archaeologists often search for burial sites, which are underground to begin with.

Really old cities were not exactly constructed with durable materials. In Mesopotamia (modern Iraq), everything was basically made of mud. These mud buildings fell apart quite easily, and the next person who came along simply built on top of the rubble. This went on for ages, until there was a nice thick layer of muddy mush with the occasional surviving piece of detail.

Ancient standards of cleanliness didn't help, either. Rubbish piled up around the houses, contributing to the mush. Eventually, people would leave (and really, who could blame them?), and their shoddy mud houses would get reduced to a muddy mound, called a tell. Now, archaeologists look for tells and dig down into them. At the bottom of the mounds of dirt, they can often find the remains of buildings like temples. Temples tended to survive because they were kept clean and in good repair. They often incorporated stone elements into their construction, making them more durable.

Ancient cities that were built mostly of better materials like stone (Rome, for example) tend not to be completely buried. Sure, they're buried a bit, but that's due to the things we mentioned before: growth of vegetation and rubbish piling up. Trouble is, stone robbers came and swiped the ancient rock to make their own buildings, leaving many stone sites in a sad state.

The moral of the story? At least underground ruins are kept safe—until someone discovers them.

Q Who were the Druids and the Picts?

A What do you know about the Druids? How about the Picts? Chances are, what you know (or think you know) is wrong. These two "lost" peoples are saddled with serious cases of mistaken identity. Most contemporary perceptions of the Druids and Picts tend to be derived from legend and lore. As such, our conceptions of these peoples range from erroneous and unlikely to just plain foolish.

Let's start with the Druids. They are often credited with the building of Stonehenge, the great stone megalith believed to be their sacred temple, as well as their arena for savage human sacrifice rituals. True or False? False. First of all, Stonehenge was built around 2000 B.C.—1,400 years before the Druids emerged. Second, though we know admittedly little of Druidic practice, it seemed to be traditional and conservative. The Druids did have specific divinity-related beliefs, but it is not known whether they actually carried out human sacrifices.

As the priestly class of Celtic society, the Druids served as the Celts' spiritual leaders—repositories of knowledge about the world and the universe, as well as authorities on Celtic history, law, religion, and culture. In short, they were the preservers of the Celtic way of life.

The Druids provided the Celts with a connection to their gods, the universe, and the natural order. They preached of the power and authority of the deities and taught the immortality of the soul and reincarnation. They were experts in astronomy and the natural world. They also had an innate connection to all things living: They preferred holding great rituals among natural shrines provided by the forests, springs, and groves.

To become a Druid, one had to survive extensive training. Druid wannabes and Druid-trained minstrels and bards had to endure as many as 20 years of oral education and memorization.

In terms of power, the Druids took a backseat to no one. Even the Celtic chieftains, well-versed in power politics, recognized the overarching authority of the Druids. Celtic society had well-defined power and social structures and territories and property rights. The Druids were deemed the ultimate arbiters in all matters relating to such. If there was a legal or financial dispute between two parties, it was unequivocally settled in special Druid-presided courts. Armed conflicts were immediately ended by Druid rulings. Their word was final.

In the end, however, there were two forces to which even the Druids had to succumb—the Romans and Christianity. With the Roman invasion of Britain in A.D. 43, Emperor Claudius decreed that Druidism throughout the Roman Empire was to be outlawed. The Romans destroyed the last vestiges of official Druidism in Britain with the annihilation of the Druid stronghold of Anglesey in A.D. 61. Surviving Druids fled to unconquered Ireland and Scotland, only to become completely marginalized by the influence of Christianity within a few centuries.

Stripped of power and status, the Druids of ancient Celtic society disappeared. They morphed into wandering poets and storytellers with no connection to their once illustrious past.

What about the Picts? Although often reduced to a mythical race of magical fairies, the Picts actually ruled Scotland before the Scots. Their origins are unknown, but some scholars believe that the Picts were descendants of the Caledonians or other Iron Age tribes who invaded Britain.

No one knows what the Picts called themselves; the origin of their name comes from other sources and probably derives from the Pictish custom of tattooing or painting their bodies. The Irish called them Cruithni, meaning "the people of the designs." The Romans called them Picti, which is Latin for "painted people"; however, the Romans probably used the term as a general moniker for all the untamed peoples living north of Hadrian's Wall.

The Picts themselves left no written records. All descriptions of their history and culture come from second-hand accounts. The earliest of these is a Roman account from A.D. 297 stating that the

Picti and the Hiberni (Irish) were already well-established enemies of the Britons to the south.

The Picts were also well-established enemies of each other. Before the arrival of the Romans, the Picts spent most of their time fighting amongst themselves. The threat posed by the Roman conquest of Britain forced the squabbling Pict kingdoms to come together and eventually evolve into the nation-state of Pictland. The united Picts were strong enough not only to resist conquest by the Romans, but also to launch periodic raids on Roman-occupied Britain.

Having defied the Romans, the Picts later succumbed to a more benevolent invasion launched by Irish Christian missionaries. Arriving in Pictland in the late 6th century, they succeeded in converting the polytheistic Pict elite within two decades. Much of the written history of the Picts comes from the Irish Christian annals. If not for the writings of the Romans and the Irish missionaries, we might not have knowledge of the Picts today.

Despite the existence of an established Pict state, Pictland disappeared with the changing of its name to the Kingdom of Alba in A.D. 843, a move signifying the rise of the Gaels as the dominant people in Scotland. By the 11th century, virtually all vestiges of them had vanished.

Q Who found the Dead Sea Scrolls?

A While rounding up a stray animal near Qumran, Israel, in early 1947, Bedouin shepherd Mohammed el-Hamed stumbled across several pottery jars containing scrolls written in Hebrew. It turned out to be the find of a lifetime.

News of the exciting discovery of ancient artifacts spurred archaeologists to scour the area of the original find for additional material. Over a period of nine years, the remains of approximately 900 documents were recovered from 11 caves near the ruins of Qumran, a plateau community on the northwest shore of the Dead Sea. The documents have come to be known as the Dead Sea Scrolls.

Tests indicate that all but one of the documents were created between the middle of the 2nd century B.C. and the 1st century A.D. Nearly all were written in one of three Hebrew dialects. The majority of the documents were written on animal hide.

The scrolls represent the earliest surviving copies of Biblical documents. Approximately 30 percent of the material is from the Hebrew Bible. Every book of the Old Testament is represented with the exception of the Book of Esther and the Book of Nehemiah. Another 30 percent of the scrolls contain essays on subjects including blessings, war, community rule, and the membership requirements of a Jewish sect. About 25 percent of the material refers to Israelite religious texts not contained in the Hebrew Bible, while 15 percent of the data has yet to be identified.

Since their discovery, debate about the meaning of the scrolls has been intense. One widely held theory subscribes to the belief that the scrolls were created at the village of Qumran and then hidden by the inhabitants. According to this theory, a Jewish sect known as the Essenes wrote the scrolls. Those subscribing to this theory have concluded that the Essenes hid the scrolls in nearby caves during the Jewish Revolt in A.D. 66, shortly before they were massacred by Roman troops.

A second major theory, put forward by Norman Golb, Professor of Jewish History at the University of Chicago, speculates that the scrolls were originally housed in various Jerusalem-area libraries and were spirited out of the city when the Romans besieged the capital in A.D. 68–70. Golb believes that the treasures documented on the so-called Copper Scroll could only have been held in Jerusalem. Golb also alleges that the variety of conflicting ideas found in the scrolls indicates that the documents are facsimiles of literary texts.

The documents were catalogued according to which cave they were found in and have been categorized into Biblical and non-Biblical works. Of the eleven caves, numbers 1 and 11 yielded the most intact documents, while number 4 held the most material—an astounding 15,000 fragments representing 40 percent of the total material found. Multiple copies of the Hebrew Bible have been identified, including 19 copies of the Book of Isaiah, 30 copies of Psalms, and 25 copies of Deuteronomy. Also found were previously unknown psalms attributed to King David, and stories about Abraham and Noah.

Most of the fragments appeared in print between 1950 and 1965, with the exception of the material from Cave 4. Publication of the manuscripts was entrusted to an international group led by Father Roland de Vaux of the Dominican Order in Jerusalem.

Access to the material was governed by a "secrecy rule"—only members of the international team were allowed to see them. In late 1971, 17 documents were published, followed by the release of a complete set of images of all the Cave 4 material. The secrecy rule was eventually lifted, and copies of all documents were in print by 1995.

Many of the documents are now housed in the Shrine of the Book, a wing of the Israel Museum located in Western Jerusalem. The scrolls on display are rotated every three to six months.

Q What was the Leichhardt expedition?

A Ludwig Leichhardt, a Prussian linguist and naturalist, traveled to Australia in his late 20s in search of knowledge and adventure. He spent two years collecting and studying geological and botanical samples in the area around Sydney. In August 1844, he embarked upon a privately funded expedition of the northeast coast of the continent. Since his party of ten had long been given up for dead, they were hailed as heroes when they arrived in Port Essington on December 17, 1845.

Leichhardt next planned a government-sponsored expedition to traverse the entire continent from east to west. Setting out in December 1846, the group struggled against heavy rains, bouts of malaria, and a lack of food. They returned seven months later, having covered only 500 miles. Undaunted, Leichhardt mounted another attempt to cross the continent in March 1848. The party stopped briefly at an outpost known as McPherson's Station in April and then continued on their way. They were never heard from again.

Over the next ten years, several attempts to find Leichhardt were made, but little evidence was found. The fate of the expedition has remained one of the great mysteries of Australian exploration. Around 1900, a brass strip bearing Leichhardt's name and the date 1848 was discovered by an Aborigine named Jackie. The strip was attached to a burnt-out firearm. Jackie gave the gun to his boss, prospector Charles Harding, who threw the gun away but treasured the nameplate. Harding kept the plate hidden but would occasionally show it to friends.

Around 1917, Harding gave the plate to his friend Reginald Bristow-Smith. Bristow-Smith loaned the plate to the South

Australian Museum in 1920 and to explorer Larry Wells in 1934. Wells let the Royal Geographical Society of Australia study the plate, and in 1937, the society declared the plate genuine. Government agencies then lost track of the plate, and it was not returned to Bristow-Smith until 1964. Historian Dr. Darrell Lewis helped the National Museum of Australia get in contact with the Bristow-Smith family in 2005, and the museum purchased the plate in 2006. Because Jackie discovered the plate in western Australia, the nameplate shows that some members of the expedition likely made it at least two-thirds of the way to their destination.

Q **What are some of the world's greatest missing treasures?**

A They were fantastic examples of opulence, decadence, and splendor. People marveled at their beauty, drooled over their excess, and cowered at their power. And now they're gone. But where are they? And what happened to them?

1. The Amber Room

Described as the eighth wonder of the world by those who saw it, the Amber Room is certainly the most unique missing treasure in history. It was an 11-foot-square hall consisting of large wall panels inlaid with several tons of superbly designed amber, large gold-leaf-edged mirrors, and four magnificent Florentine mosaics. Arranged in three tiers, the amber was inlaid with precious jewels, and glass display cases housed one of the most valuable collections of Prussian and Russian artwork ever assembled. Created for Prussia's King Friedrich I, and given to Russian czar Peter the Great in 1716, it was located at Catherine Palace, near

St. Petersburg. Today, the Amber Room would be valued at more than $142 million.

When Adolf Hitler turned his Nazi war machine toward Russia, the keepers of the Amber Room got nervous. They tried to move it, but the amber began to crumble, so they tried to cover it with wallpaper. They were unsuccessful and when the Nazis stormed Leningrad (formerly called St. Petersburg) in October 1941, they claimed it and put it on display in Konigsberg Castle during the remaining war years. But, when Konigsberg surrendered in April 1945, the fabled treasure was nowhere to be found. The Amber Room was never seen again. Did the Soviets unwittingly destroy their own treasure with bombs? Was it hidden in a now lost subterranean bunker outside the city? Or was it destroyed when Konigsberg Castle burned shortly after the city surrendered? We'll probably never know for sure. But fortunately for lovers of opulence, the Amber Room has been painstakingly recreated and is on display in Catherine Palace.

2. Blackbeard's Treasure

The famous pirate Blackbeard only spent about two years (1716-1718) plundering the high seas. Within that time, however, he amassed some serious wealth. While the Spanish were busy obtaining all the gold and silver they could extract from Mexico and South America, Blackbeard and his mates waited patiently in the waters, then pounced on the treasure-laden ships as they sailed back to Spain.

Blackbeard developed a fearsome reputation as a cruel and vicious opportunist. His reign of terror centered around the West Indies and the Atlantic coast of North America, with headquarters in both the Bahamas and North Carolina. His end came in November 1718, when British Lieutenant Robert Maynard decapitated the pirate and hung his head from the bowsprit of his ship as a grisly trophy.

But what happened to the vast treasure that Blackbeard had amassed? He acknowledged burying it but never disclosed the location. But that hasn't stopped countless treasure hunters from trying to get their hands on it. Blackbeard's sunken ship, *Queen Anne's Revenge,* is believed to have been discovered near Beaufort, North Carolina, in 1996, but the loot wasn't onboard. Possible locations for the hidden stash include the Caribbean Islands, Virginia's Chesapeake Bay, and the caves of the Cayman Islands.

3. Treasures of Lima

In 1820, Lima, Peru, was on the edge of revolt. As a preventative measure, the viceroy of Lima decided to transport the city's fabulous wealth to Mexico for safekeeping. The treasures included jeweled stones, candlesticks, and two life-size solid gold statues of Mary holding the baby Jesus. In all, the treasure filled 11 ships and was valued at around $60 million.

Captain William Thompson, commander of the *Mary Dear,* was put in charge of transporting the riches to Mexico. But the viceroy should have done some research on the man to whom he handed such fabulous wealth because Thompson was a pirate, and a ruthless one at that. Once the ships were well out to sea, he cut the throats of the Peruvian guards and threw their bodies overboard.

Thompson headed for the Cocos Islands, in the Indian Ocean, where he and his men allegedly buried the treasure. They then decided to split up and lay low until the situation had calmed down, at which time they would reconvene to divvy up the spoils. But the *Mary Dear* was captured, and the crew went on trial for piracy. All but Thompson and his first mate were hanged. To save their lives, the two agreed to lead the Spanish to the stolen treasure. They took them as far as the Cocos Islands and then managed to escape into the jungle. Thompson, the first mate, and the treasure were never seen again.

4. Pharaohs' Missing Treasure

When Howard Carter found the tomb of Tutankhamen in Egypt's Valley of the Kings in 1922, he was mesmerized by the splendor of the artifacts that the young king took to the afterlife. Attached to the burial chamber was a treasury with so many jewels and other artifacts that it took Carter ten years to fully catalog them. However, when the burial chambers of more prominent pharaohs were unearthed in the late 19th century, their treasure chambers were virtually empty. It is common knowledge that tomb robbers had been busy in the tombs over the centuries, but the scale of the theft required to clean out the tombs of the kings is beyond petty criminals. So, where is the vast wealth of the pharaohs buried in the Valley of the Kings?

Some scholars believe that the treasures were appropriated by the priests who conducted reburials in the Valley of the Kings during the period of the early 20th and late 21st Egyptian dynasties (425–343 B.C.). Pharaohs were not averse to reusing the funeral splendors of their ancestors, so this may have been carried out with official sanction. One particular ruler, Herihor, has been the focus of special attention. Herihor was a high court official during the reign of Ramses XI. Upon Ramses' death, Herihor usurped the throne, dividing up the kingdom with a co-conspirator, his son-in-law Piankh. Herihor placed himself in charge of reburial proceedings at the Valley of the Kings, affording himself ample opportunity to pilfer on a grand scale. His tomb has never been found. When and if it is, many scholars believe that the missing treasures of many of Egypt's pharaohs will finally see the light of day.

5. Montezuma's Treasure

The Spanish decimation of the Aztec empire in Mexico came to a head on July 1, 1520. After mortally wounding Emperor Montezuma, Hernando Cortés and his men were besieged by enraged Aztec warriors in the capital city of Tenochtitlán. After days of fierce fighting, Cortés ordered his men to pack up the vast treasures of Montezuma in preparation for a night flight, but they didn't get far before the Aztecs fell upon them. The ensuing carnage filled Lake Tezcuco with Spanish bodies and the stolen treasures of Montezuma. The terrified army had thrown the booty away in a vain attempt to escape with their lives. The hoard consisted of countless gold and silver ornaments, along with a huge array of jewels.

Cortés and a handful of his men got away with their lives and returned a year later to exact their revenge. When the inhabitants of Tenochtitlán got wind of the approaching invaders, they buried the remains of the city's treasure in and around Lake Tezcuco to prevent it from falling prey to the gold-crazed Spanish. Today, a vast treasure trove remains hidden beneath nearly five centuries of mud and sludge on the outskirts of Mexico City, the modern day incarnation of Tenochtitlán. Generations of treasure seekers have sought the lost hoard without success. A former president of Mexico even had the lake bed dredged, but no treasure was found.

Q What's become of the most famous ancient cities?

A In the ancient world, it took far fewer people to make a great city. Some didn't survive; some have flourished; and others have exploded. With the understanding that ancient population estimates are necessarily approximate, here are the fates of some great metropolises:

Memphis (now the ruins of Memphis, Egypt): By 3100 B.C., this Pharaonic capital bustled with an estimated 30,000 people. Today it has none—but modern Cairo, 12 miles north, houses more than 7 million people.

Ur (now the ruins of Ur, Iraq): Sumer's great ancient city once stood near the Euphrates with a peak population of 65,000 around 2030 B.C. Now its population is zero, and the Euphrates has meandered about ten miles northeast.

Alexandria (now El-Iskandariya, Egypt): Built on an ancient Egyptian village site near the Nile Delta's west end, Alexander the Great's city once held a tremendous library. In its heyday, it may have held 250,000 people; today more than 3,300,000 people call it home.

Babylon (now the ruins of Babylon, Iraq): Babylon may have twice been the largest city in the world, in about 1700 B.C. and 500 B.C.—perhaps with up to 200,000 people in the latter case. Now, it's windblown dust and faded splendor.

Athens (Greece): In classical times, this powerful city-state stood miles from the coast but was never a big place—something like 30,000 residents during the 300s B.C. It now reaches the sea with about 3,000,000 residents.

Rome (Italy): With the rise of its empire, ancient Rome became a city of more than 500,000 and the center of Western civilization. Though that mantle moved on to other cities, Rome now has 3,000,000 people.

Constantinople (now Istanbul, Turkey): First colonized by Greeks in the 1200s B.C., this city of fame was made Emperor Constantine the Great's eastern imperial Roman capital with

300,000 people. As Byzantium, it bobbed and wove through the tides of faith and conquest. Today, it is Turkey's largest city with 10,000,000 people.

Xi'an (China): This longtime dynastic capital, famed for its terra-cotta warriors but home to numerous other antiquities, reached 400,000 people by A.D. 637. Its nearly 8,000,000 people make it as important a city now as then.

Baghdad (Iraq): Founded around A.D. 762, this center of Islamic culture and faith was perhaps the first city to house more than 1,000,000 people. It has sometimes faded but never fallen.

Carthage (now the ruins of Carthage, Tunisia): Phoenician seafarers from the Levant founded this great trade city in 814 B.C. Before the Romans obliterated it in 146 B.C., its population may have reached 700,000. Today, it sits in empty silence ten miles from modern Tunis—population 2,000,000.

Tenochtitlán (now Mexico City, Mexico): Founded in A.D. 1325, this island-built Aztec capital had more than 200,000 inhabitants within a century. Most of the surrounding lake has been drained over the years. A staggering 19,000,000 souls call modern Mexico City home.

Q What happens to all the stuff we launch into space and don't bring back?

A Space trash creates a major traffic hazard. If you think it's nerve-wracking when you have to swerve around a huge pothole as you cruise down the highway, just imagine how it would feel if you were hundreds of miles above the surface of Earth, where the stakes couldn't be higher. That's the situation that the crew of the International Space Station (ISS) faced in 2008 when it had to perform evasive maneuvers to avoid debris from a Russian satellite.

And that was just one piece of orbital trash—all in all, there are tens of millions of junky objects that are larger than a millimeter and are in orbit. If you don't find this worrisome, imagine the little buggers zipping along at up to seventeen thousand miles per hour. Worse, these bits of flotsam and jetsam constantly crash into each other and shatter into even more pieces.

The junk largely comes from satellites that explode or disintegrate; it also includes the upper stages of launch vehicles, burnt-out rocket casings, old payloads and experiments, bolts, wire clusters, slag and dust from solid rocket motors, batteries, droplets of leftover fuel and high-pressure fluids, and even a space suit. (No, there wasn't an astronaut who came home naked—the suit was packed with batteries and sensors and was set adrift in 2006 so that scientists could find out how quickly a spacesuit deteriorates in the intense conditions of space.)

So who's responsible for all this orbiting garbage? The two biggest offenders are Russia—including the former Soviet Union—and the United States. Other litterers include China, France, Japan, India, Portugal, Egypt, and Chile. Each of the last three countries has launched one satellite during the past twenty years.

Most of the junk orbits Earth at between 525 and 930 miles from the surface. The Space Shuttle and the ISS operate a little closer to Earth—the Shuttle flies at between 250 and 375 miles up, and the ISS maintains an altitude of about 250 miles—so they don't see the worst of it. Still, the ISS's emergency maneuver in 2008 was a sign that the situation is getting worse. Houston, we have a problem.

NASA and other agencies use radar to track the junk and are studying ways to get rid of it for good. Ideas such as shooting at objects with lasers or attaching tethers to some pieces to force them back to Earth have been discarded because of cost considerations and the potential danger to people on the ground. Until an answer is found, NASA practices constant vigilance, monitoring the junk and watching for collisions with working satellites and vehicles as they careen through space. Hazardous driving conditions, it seems, extend well beyond Earth's atmosphere.

Q Where was Petra?

A In the wilds of southern Jordan lies one of antiquity's most beautifully preserved sights: what survives of Petra, the fabulous "Red City" that was the ancient Nabataean capital.

Petra lies within the Hashemite Kingdom of Jordan, perhaps 80 miles south of Amman in the Naqab Desert, about 15 miles east of the Israeli border. It is a World Heritage Site and a Jordanian national treasure, cared for accordingly. Once, Petra was a key link in the trade chain connecting Egypt, Babylon, Arabia, and the Mediterranean. Despite being in the desert, it had water (if you knew how to look) and was quite defensible.

In 600 B.C., the narrow red sandstone canyon of Petra housed a settlement of Edomites: seminomadic Semites said to descend from Biblical Esau. Egypt was still rich but declining. Rome was a young farming community dominated by its Etruscan kings. The rise of classical Athens was decades away. Brutal Assyria had fallen to Babylonian conquerors. With the rise of the incense trade, Arab traders began pitching tents at what would become Petra. We know them as the Nabataeans.

Nabataean history spanned a millennium. They showed up speaking early Arabic in a region where Aramaic was the business-speak. The newcomers thus first wrote their Arabic in a variant of the Aramaic script. But Petra's trade focus meant a need to adopt Aramaic as well, so Nabataeans did—many words crossed the Arabic/Aramaic linguistic fence at Petra. By the end (about 250 years before the rise of Islam), Nabataean "Arabaic" had evolved into classical (Koranic) Arabic.

The Nabateans weren't expansionists, but defended their homeland with shrewd diplomacy and obstinate vigor. Despite great wealth, they had few slaves. Despite monarchical government, Petra's Nabataeans showed a pronounced democratic streak. Empires rose and fell around them; business was business.

And their trade was very lucrative. The core commodity was incense from Arabia, but many raw materials and luxuries of antiquity also passed through Petra—notably bitumen (natural asphalt), useful in waterproofing and possibly in embalming.

Petra's heyday was around 70 B.C., when it had a population of about 20,000. If you'd walked the streets then, you'd have seen ornate homes and public buildings rivaling Athenian and Roman artistry, all carved into the high red sandstone walls of the canyon. You'd see a camel caravan arriving from Arabia loaded with goods and white-robed traders dismounting with elegant gifts for their buying contacts. The wealthy aroma of frankincense would constantly remind your nostrils why Petra existed. Most people wore robes and cloaks, often colored by exotic dyes. Petra was luxurious without being licentious.

You'd overhear conversations in Aramaic and Arabic: A new cistern was under construction in the nearby hills. Workers were shoring up a building damaged by a recent earth tremor. Old-timers groused that reigning King Aretas III wished he were

Greek. A modestly robed vendor walked past with dates for sale; you'd fish out a thick silver coin to offer her. Along with your bronze change and the delicious dates, she would wish you the favor of al-Uzza, the Nabataean goddess identified with Aphrodite and Venus.

If you asked a passing water-bearer about that guy in the outlandish robe draped over one shoulder, followed by servants, you'd hear that he was a man of faraway Rome. At the time, Rome was a dynamic market for Petra's goods, with domains beginning to rival Alexander the Great's once-mighty empire. Only time would tell how Petra will reckon with this next tide of power.

Petra's last king, Rabbel II, willed his realm to Rome. When he died in A.D. 106, Nabataea became the Roman province of Arabia Petraea. Again the Nabataeans adjusted—and kept up the trade. In the 2nd and 3rd centuries A.D., the caravans began using Palmyra (in modern Syria) as an alternate route, starting a long, slow decline at Petra. An earthquake in 363 delivered the knockout punch: damage to the intricate water system sustaining the city. By about 400, Petra was an Arabian ghost town.

But it is still visited today as a historical site, by thousands of people each day. If you can travel to Jordan, you can travel to Petra—either with an organized tour booked through a travel agent or on your own if that's your style. Nearby hotels and restaurants offer modern accommodations. The site charges a daily entrance fee.